Biobanking in the Era of the Stem Cell: A Technical and Operational Guide

Colloquium Series on
Stem Cell Biology

Editors

Wenbin Deng, Ph.D., *Department of Cell Biology and Human Anatomy, Institute for Pediatric Regenerative Medicine, School of Medicine, University of California, Davis*

This series is interested in covering the fundamental mechanisms of stem cell pluripotency and differentiation, and the strategies of translating fundamental developmental insights into discovery of new therapies. The emphasis is on the roles and potential advantages of stem cells in developing, sustaining and restoring tissue after injury or disease. Some of the topics included will be the signaling mechanisms of development and disease; the fundamentals of stem cell growth and differentiation; the utilities of adult (somatic) stem cells, induced pluripotent stem (iPS) cells and human embryonic stem (ES) cells for disease modeling and drug discovery; and the prospects for applying the unique aspects of stem cells for regenerative medicine. We hope this series will provide the most accessible and current discussions of the key points and concepts in the field, and that students and researchers all over the world will find these in-depth reviews to be useful.

Published titles

(for future titles please see the website, www.morganclaypool.com/page/lifesci)

Biobanking in the Era of the Stem Cell: A Technical and Operational Guide
Jennifer C. Moore, Michael H. Sheldon, and Ronald P. Hart
www.morganclaypool.com

ISBN: 9781615044726 paperback

ISBN: 9781615044733 ebook

DOI: 10.4199/C00059ED1V01Y201206SCB002

A Publication in the

COLLOQUIUM SERIES ON STEM CELL BIOLOGY

Lecture #2

Series Editor: Wenbin Deng, University of California

Series ISSN Pending

Biobanking in the Era of the Stem Cell: A Technical and Operational Guide

Jennifer C. Moore
Department of Genetics and NIMH Stem Cell Center at the Rutgers University Cell & DNA Repository

Michael H. Sheldon
Department of Genetics and NIMH Stem Cell Center at the Rutgers University Cell & DNA Repository

Ronald P. Hart
Department of Cell Biology & Neuroscience, Rutgers University, Piscataway, NJ

COLLOQUIUM SERIES ON STEM CELL BIOLOGY #2

MORGAN&CLAYPOOL LIFE SCIENCES

ABSTRACT

The study of mental health disorders and the genetics behind these disorders can be greatly enhanced by the use of induced pluripotent stem cells (iPSC). Since many mental health disorders develop after puberty, the only way in which to study the genetic mechanism of these diseases previously was through cellular surrogates, such as blood or cultured fibroblasts. Having the ability to reprogram adult cells to the pluripotent stage provides the capacity to study the onset of these disorders during a culture model of neural development and to include the impact of genetic risk factors and potential environmental triggers. Working with the National Institute of Mental Health (NIMH), the Rutgers Cell and DNA Repository (RUCDR) has begun banking iPSC source cells and converting those source cells into iPSC for distribution to the scientific community. Although initial protocols were developed to reprogram fibroblasts, the ability to reprogram blood cells has several advantages including less invasive collection, less post collection manipulation, and the large number of samples in existing collections. Here, we provide detailed protocols for reprogramming either fibroblasts with retroviral vectors or cryopreserved lymphocytes with Sendai viral vectors. Our goal is to support the discovery of effective treatments for mental health disorders.

KEYWORDS

iPSC, reprogramming, mental health, genetics, pluripotency, protocols, viral vectors, fibroblasts, lymphocytes

Contents

CONTENTS

Acknowledgments

This work was supported by NIMH U24 MH068457, NIDA R21 DA032984, and the Human Genetics Institute of New Jersey. The authors thank Alana Toro-Ramos, Navindra Sukhram, Poonam Verma, Diane Huynh, and Yuming Li for development of protocols. We also thank Jay Tischfield for support, encouragement, and vision.

CHAPTER 1

Introduction

1.1   iPSC FOR THE STUDY OF MENTAL HEALTH DISORDERS

In the few years since discovery of cellular reprogramming to pluripotency, a large and robust community of researchers has begun to exploit this technology to attack many types of human disease. One of these applications, the study of human mental health disorders, is particularly suited to the use of induced pluripotent stem cells (iPSC) because there is a strong history of genetic studies and because it is nearly impossible to obtain functional tissue samples from affected subjects. Strategically, it should be productive to combine the knowledge of risk markers derived from genetic studies with the capability to produce electrically active neurons in culture for testing proposed mechanisms, screening candidate compounds as therapies, and comparing different risk alleles for unique features in understanding specific disorders. This book will review recent successes with iPSC in the study of mental health disorders and then provide specific protocols that can be used in a repository environment. Our focus on a repository is based on our involvement with a large biobank and the need to consider standardized protocols and reproducible results for distribution of cells to many research labs.

The utility of stem cells in combination with protocols for differentiating cells into neurons was recognized well before the widespread use of embryonic stem cells (ESC) or iPSC (McKay, 2004). Quickly after iPSC methods first became available (Takahashi et al., 2007; Yu et al., 2007), researchers announced applications related to neurodegeneration (Wernig et al., 2008) and genetic disorders (Ebert et al., 2009; Lee et al., 2009). This has now led to the production of model cells from mental health disorders including Rett syndrome (Marchetto et al., 2010), Timothy syndrome (Pasca et al., 2011), schizophrenia (Pedrosa et al., 2011), and autism (Derosa et al., 2012).

As an example, Rett syndrome (RTT) is an autism spectrum disorder that is caused by mutations in a methyl CpG binding protein that helps to regulate DNA methylation (MeCP2). RTT primarily affects females since MeCP2 is encoded by the X chromosome, and females randomly inactivate one X chromosome during development, leading to an effectively mosaic individual. In RTT, some cells express wild-type MeCP2 and some lack MeCP2 or express mutated versions. However, RTT represents one of the only autism-related disorders that is known to be the result of a single gene defect. Therefore, RTT is an ideal case study for demonstrating the value of iPSC in studying neurodevelopmental disorders, particularly because the differentiation of neurons from pluripotent cells in culture recapitulates the developmental process.

Human genetic diseases may be influenced by epigenetic status and RTT provides a good example of this as well. Epigenetics considers mechanisms that are preserved over cell division but are not encoded in DNA sequence. Will the life histories of subjects be reflected in persistent epigenetic patterns carried through into iPSC or will these be lost upon reprogramming? The most common mechanisms are DNA methylation and chromatin modification, but in mammals, a prominent mechanism is the random inactivation of one X chromosome in females. One study claims that iPSC reprogramming retains more epigenetic history than stem cells constructed by nuclear transfer (Kim et al., 2010). This may turn out to be valuable in, for example, comparing two individuals treated with different drug therapies but possessing the same risk SNPs for a neuropsychiatric disorder. It is important to keep these issues in mind when designing iPSC studies. In one of the first CNS disease-related iPSC studies, RTT patient fibroblasts were used to create iPSC that were capable of producing neurons in culture (Marchetto et al., 2010). Because of the X-linked nature of RTT, there was a concern that the iPSC might retain the epigenetic property of X-inactivation. This would cause problems if, for example, a cell from affected individuals, who are mosaic females, retained the inactivated allele of the mutated MeCP2 in a particular fibroblast source cell. If X-inactivation was not reversed during reprogramming, this clone of iPSC would express only the wild-type MeCP2 allele. Marchetto and colleagues found that some, but not all, iPSC clones exhibited evidence of X-reactivation and therefore only reactivated clones were studied. Once this issue was resolved, it was found that the resulting cultured RTT neurons exhibit smaller cell soma, reduced spine densities, altered calcium signaling, and electrophysiological defects (Marchetto et al., 2010), representing cellular and physiological properties that are thought to underlie RTT.

Other studies have also observed differences in the X-inactivation status of RTT iPSC, as reviewed by Cheung (2012). This demonstrates that both mutational and epigenetic issues must be considered when using human samples for iPSC modeling. Some groups utilized the unique nature of X-linked genetics to build sets of isogenic pairs of iPSC from the same patient (Ananiev et al., 2011; Cheung et al., 2011). By selecting multiple clones of iPSC from the same culture of mosaic fibroblasts, both affected and unaffected iPSC lines could be derived, differing only in the choice of which X allele is inactivated. Other examples support the role of epigenetics in mental health disorders. Schizophrenia is believed to be influenced by a number of environmental factors, including antipsychotic drug history (Melas et al., 2012). It is clear that epigenetic status is an important consideration for modeling human disease with iPSC.

All groups found that neuronal soma size was reduced in RTT iPSC-derived neurons. More advanced functional studies with mouse iPSC models show that RTT neurons show affected action potentials and inward currents pointing to a role for deficient sodium channel function in RTT (Farra et al., 2012). By focusing on both morphological properties as well as neuronal electrophysiology, disease mechanisms can be studied in human cell cultures derived from subjects with known histories. This shows that cells carrying mutations in disorder-associated genes can develop phenotypic properties in culture that are either characteristic or predictive of disease. The resulting neuronal cultures can be used for traditional mechanistic studies in research environments.

Disease-specific iPSC can also be used for screening compounds that might reverse or alleviate symptoms (Dolmetsch and Geschwind, 2011; Inoue and Yamanaka, 2011; Kim and Jin, 2012). An early example was the screening of compounds to alleviate symptoms of familial dysautonomia (FAD) by the Studer lab (Lee et al., 2009). FAD is caused by mutation of the IKBKAP gene, causing a fatal peripheral neuropathy through loss of autonomic and sensory neurons. iPSC prepared from FAD patients demonstrated neuron-specific splicing defects of IKBKAP (Lee et al., 2009). Studer was able to use this property to screen a small list of candidate drugs for use in FAD. Of the three compounds tested, only kinetin, which has previously been shown to reduce levels of mis-spliced IKBKAP in lymphoblast cell lines (Slaugenhaupt et al., 2004), resulted in a reduction in neuron-specific IKBKAP splicing. This limited study demonstrates that human iPSC-derived neurons can be used to construct screening strategies for disease. As the list of disease-specific iPSC lines grows and cellular phenotypes are identified, it is expected that there will be a rush to apply the technology of high-throughput screening. The added benefit of using human iPSC is that cells from many individuals can be included in any screening, essentially creating a "clinical trial in a dish," where distinct genotypes might be associated with specific beneficial compounds. This exemplifies the goal of so-called personalized medicine in a practical, high-throughput strategy.

This approach is completely distinct from a regenerative medicine strategy of preparing therapeutic iPSC for alleviating symptoms, such as has been explored for Parkinson's disease (Hargus et al., 2010; Liu et al., 2011; Wernig et al., 2008). The focus on mechanisms and compound screening eliminates requirements for GLP (good laboratory practice) and other certifications that might be required for an FDA-approvable cell. Most neuropsychiatric disorders are likely to be more amenable for a mechanistic and screening approach since cellular therapy is not likely to reduce symptoms (Vaccarino et al., 2011).

1.2   BUILDING UPON A ROBUST GENETICS REPOSITORY

In recent years, a consensus has emerged among both governmental agencies and individual researchers that the full elucidation of the genetic and environmental etiologies of human disease will depend critically on enhancing our ability to openly and efficiently share resources among the scientific community. The most important of these resources, generated from large cohorts, include clinical data, both from the initial diagnosis and subsequent follow-up examinations, and biological materials isolated from affected and control individuals, such as nucleic acids, plasma proteins, and cell lines. The recognition of the importance of such centralized resources led the National Institute of Mental Health (NIMH), in October 1998, to fund the establishment of the NIMH Center for Collaborative Genomic Studies on Mental Disorders (the "Center": https://www.nimhgenetics .org/), hosted by the Rutgers University Cell & DNA Repository (RUCDR: http://www.rucdr .org/). The RUCDR has samples from more than 250,000 de-identified subjects in a variety of collections. Under the Direction of Drs. Jay Tischfield at Rutgers University and John Rice at Washington University, the Center has processed to date over 130,000 blood samples for the NIMH. Each blood sample is shipped overnight to the RUCDR from a collection site. The blood samples are processed on the day of receipt following a number of Standard Operating Procedures (SOPs), depending on the specifications of the respective projects. These SOPs include extraction of DNA, RNA, and/or plasma from whole blood, and transformation of lymphocytes with Epstein–Barr Virus (EBV) to give rise to lymphoblastoid cell lines (LCLs). DNA extracted from whole blood is a vital resource for genomic approaches such as genome-wide SNP or copy number variation (CNV) studies, Next Generation sequencing, and epigenetics studies. The LCLs are valuable as renewable sources of genetic material, as well as a source of living cells for functional studies. Large groups have already established collections of biospecimens for the study of genetic association with mental disorders and demonstrated their utility. As these selected examples illustrate, studies have identified risk loci for schizophrenia (Ripke et al., 2011), depression (Holmans et al., 2007), Tourette syndrome (Fernandez et al., 2012), autism (Sanders et al., 2011), and bipolar disorder (Sklar et al., 2011). The NIMH collection represents a unique resource for associating diagnostic data, a range of human variation, and pre-identified risk SNPs or CNVs. The resources provided by the Center have resulted in the publication of over 500 peer-reviewed papers to date (see https://www.nimhgenetics .org/publications/).

The NIMH Stem Cell Resource (http://nimhstemcells.org) was created to supplement the existing NIMH Center, with a mission to receive, store, and distribute cells for iPSC creation and to develop methods for iPSC production from those cells. Since early methods for producing iPSC did not include lymphocytes or their derivatives, we began by establishing protocols with which to reprogram fibroblasts derived from skin biopsies. In an attempt to build upon the vast resources already stored in our repository, we also worked to adapt procedures for utilizing lymphocytes as starting material. Both methods have proven to be effective and so both are covered here. Our overall goal is to maintain an effective, well-documented workflow (Fig. 1) to produce source cells and iPSC with high quality.

Several groups have developed methods for reprogramming lymphocytes into iPSC. These methods rely on lentiviral vectors (Loh et al., 2010; Staerk et al., 2010), episomal vectors (Chou et al., 2011), or Sendai viral vectors (Seki et al., 2010) to introduce ectopic expression of iPSC factors. While the bulk of our collection is stored in the form of LCLs and methods have been published to convert LCLs to iPSC, we are only in the early stages of developing protocols for these cells. During the course of isolating lymphocytes for EBV transformation, the remaining lymphocytes that are not needed for generation of an LCL are kept and stored as cryopreserved lymphocytes (CPLs) to serve as backup material. We found that methods for isolating CD4$^+$ T cells from CPLs, activating them, and then infecting with viral vectors resulted in robust production of iPSC colonies. We chose to focus on Sendai-based systems because they were non-integrating and infect lymphocytes with high efficiency. The development of highly efficient techniques for reprogramming CPLs to iPSCs, detailed in the protocols in this book, has established CPLs as a viable source cell choice for iPSC reprogramming.

In the following sections, we will summarize key points of the protocols which will then be presented in detail in the Appendix. Our goal is to provide a convenient reference for any experienced lab to create iPSC for use in the study of neural disorders, but the protocols are sufficiently general to apply to most any application of iPSC.

1.3 PROTOCOLS SUMMARY

Each protocol is written for use in the laboratory. Some reagents, formulas, or protocols are referenced repeatedly throughout the more detailed protocols so they are made easily accessible by hyperlink within the document. Our goal is to establish uniform, documented methods for generating reproducibly high-quality products. The entire workflow is summarized in Figure 1.

1.3.1 Receiving and Banking Source Cells

In our repository (RUCDR), there is a precise protocol for submitting biological samples. In the first step, members of the group responsible for an individual collection log into the laboratory information management system (LIMS) and create a record for the sample. Based on this record, a collection kit is shipped to the collection site, including detailed protocols and all required kit components. The kit includes a shipping container and an express shipping label for delivery to the RUCDR and after the sample is collected it is sent in for processing. Upon receipt, the RUCDR documents the opening of every package by photography and then a complete set of barcoded labels is printed for each sample. These labels travel with the sample throughout its processing stages. PBMC (peripheral blood monocyte cells) are prepared from whole blood by centrifugation. Aliquots are then either stored as CPLs or LCL generation is initiated by infection of the CPLs with infection by EBV. Skin biopsies are sent to the fibroblast team where they are cultured to obtain fibroblasts. Each cell type is frozen at the appropriate stage (CPL, LCL, or fibroblast) to ensure the sample is preserved in the collection.

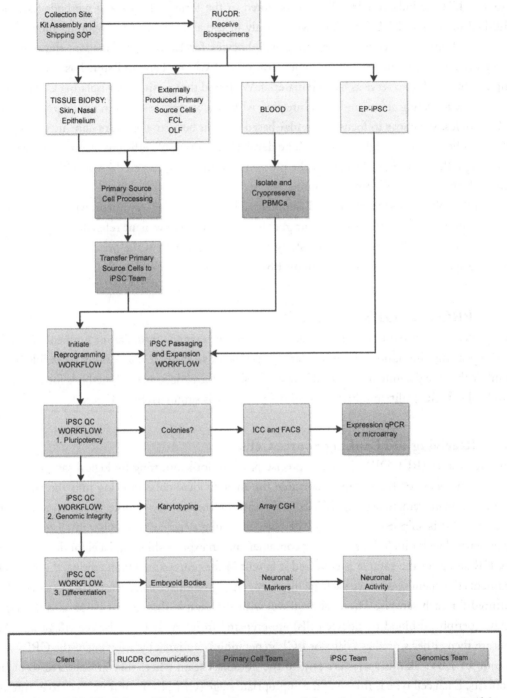

FIGURE 1: iPSC Workflow. This diagram shows the workflow that bio-specimens received by the NIMH Stem Cell Repository undergo from collection to reprogramming to characterization.

1.3.2 Preparing iPSC from Fibroblasts with Retroviral Vectors

For skin biopsies, fibroblast cultures are transferred to the stem cell team for creation of iPSC. This can be done by infecting with either retroviruses or Sendai viruses, but we will provide protocols for only retroviruses here. We prepare retroviruses from plasmid stocks (originally obtained from AddGene) in HEK293T cultures. Supernatants from these infected cells, now containing infectious virus, are added to fibroblast cultures. Six days after infection with reprogramming factors, these are then transferred to monolayers of irradiated mouse embryonic feeders (MEFs). After 14–28 days, individual colonies bearing a pluripotent phenotype are hand-picked and plated on Matrigel to expand cultures prior to freezing aliquots for distribution. Each source cell culture is generally stored as ~3–5 individual hand-picked colonies to ensure that inadvertent disruptive integration of retroviruses into host genes does not confound cell properties. All sub-lines are tested using a standard series of quality control assays, outlined below.

1.3.3 Preparing iPSC from Lymphocytes with Sendai Vectors

For CPLs, thawed lymphocytes are enriched for $CD4^+$ T cells using a negative-selection magnetic bead kit. Once T cells are selected they are activated with IL-2 and CD3/CD28 beads to promote expansion for about four days. When sufficient activated T cells have accumulated they are infected with commercially produced Sendai viral vectors containing iPSC factors. Two days following infection they are plated onto MEFs and allowed to reprogram until pluripotent colonies are visible (~14–21 days). Colonies are then hand-picked and expanded on Matrigel. Since Sendai vectors do not integrate into host chromosomes, there is not the concern for picking multiple colonies per line to prevent insertional activation. However, there may be issues with the extent of reprogramming, the robustness of individual sub-lines, or some other unforeseen rationale, so at this point, we continue to pick up to three colonies per CPL sample. As we gain experience with this cells we expect this to change. Once again, all sub-lines are subjected to quality control assays.

1.3.4 Low O_2 Culture

A few reports suggest that pluripotency is enhanced by culture in the presence of low oxygen tension (Panchision, 2009). Interestingly, while our confirmation experiments are not complete, we find roughly twice as many colonies formed from CPL with Sendai infection when we use low (4%) O_2 culture incubators, and the reprogramming of fibroblasts is also enhanced in low O_2 culture. In order to maximize our reprogramming efficiency all of our cell lines are kept at low O_2 and iPSC generation is also done at low O_2.

1.3.5 Quality Control

We have previously considered strategies for documenting pluripotency in human embryonic stem cells (Moore et al., 2010). Selection of appropriate assays is similarly important for iPSC. After careful consideration of assays used by others and assays we have used previously with hESC, we chose a thorough series of protocols, listed below.

(1) **Live cell staining.** Prior to picking colonies, the presence of a specific morphology consistent with pluripotency is usually observed. However, products are available for rapidly staining cells for the presence of alkaline phosphatase activity, an indicator of pluripotency (Singh et al., 2012). Alternatively, fluorescent antibodies can be used specifically for extracellular proteins or carbohydrate moieties, such as TRA-1-60.

(2) **Fluorescent-activated cell sorting (FACS).** While several markers have been identified for characterizing pluripotency, we use TRA-1-60 and Oct4 to test for the proportion of cells expressing both markers. Generally >95% of cells are positive for both markers.

(3) **Embryoid body (EB) formation.** The most convincing proof of pluripotency is the formation of a teratoma tumor in immunocompromised mice. While effective, this assay is time-consuming and expensive to run. As many other groups have already done, we have decided to rely on a simple in vitro differentiation assay where EBs are assayed with immunocytochemistry to detect smooth muscle actin (SMA, mesoderm), β3-tubulin (TuJ1; ectoderm) and α-fetoprotein (AFP, endoderm).

(4) **Gene expression by qPCR.** We previously used a commercial product for assessing patterns of pluripotent gene expression (Moore et al., 2010), but this turned out to be too expensive for routine use. We developed a 96-well assay using the Roche Universal Probe Library to detect many of the same target genes. Since our primary interest is in differentiation into neurons, we also expanded the list of target genes that are characteristic of various neuronal subtypes, receptors, and neurotransmitters. After collecting the cycle threshold (C_t) values for each sample and gene, a ΔC_t is calculated by subtracting the C_t for an endogenous control gene such as GAPDH or ACTB. This ΔC_t, representing a log value proportional to abundance, is then hierarchically clustered with standards (hESC, source cells) using correlation as the metric. Results generally indicate the clustering of pluripotent cells together, distinct from differentiated source cells and product cells such as neurons or neural stem cells.

(5) **Karyotyping.** To ensure that lines have not induced gross chromosomal abnormalities we run a standard G-banding karyotype. Since most of our source cells have already been studied by genome-wide association study (GWAS) and/or chromosomal abnormality, this is more of a confirmation that the iPSC protocol did not introduce new rearrangements than a test based on the source subject genotype.

(6) **Genomic integrity by CNV.** Concern has been raised about smaller genomic variations that may be induced during iPSC formation (Gore et al., 2011; Hussein et al., 2011).

These are generally too small to detect by karyotyping. We prepare DNA from the iPSC as well as from the source cells and send them for Agilent CNV Microarray assay. Results are considered to be acceptable if they are within the range of variation observed previously (Gore et al., 2011; Hussein et al., 2011).

(7) **Mycoplasma.** Routine mycoplasma testing is done by qPCR. This method provides several advantages. In addition to the high level of sensitivity associated with PCR assays, it has the ability to detect a broad spectrum of species of Mycoplasma, and is amenable to high throughput screening using laboratory automation. The current PCR assay is capable of detecting less than 10 copies of mycoplasma DNA from a 100-μl supernatant sample.

(8) **RUID™.** The RUCDR has developed a 96-SNP assay, run on a Fluidigm platform by qPCR, which establishes a unique identity of every cell sample in the repository. This assay, termed "RUID™," confirms that iPSC lines can be traced directly to the source repository specimens already stored in our LIMS. It also reports any instance of culture-to-culture contamination. This is a routine, powerful tool in maintaining a large and busy repository operation.

(9) **Sendai assays.** For iPSC derived from Sendai vectors, we wished to determine when cultures lose their Sendai genomes by attrition. The commercial vectors have been engineered to be slightly defective at replication at normal culture temperatures (Seki et al., 2010). We have incorporated a simple qPCR assay to detect Sendai genomic RNA into our methods. In preliminary results, the level of Sendai genomic RNA decreases over 1,000-fold from passage 2 to passage 10, and by passage 13 the cultures are indistinguishable from uninfected cultures. Our goal will be to document that iPSC cultures are devoid of Sendai genome upon distribution.

(10) **Methylation.** To demonstrate reprogramming of endogenous transcription factor genes, we assess CpG methylation status within promoter regions for selected genes. Standard methods require bisulfite conversion of non-methylated C residues followed by PCR amplification, cloning and sequencing (Park et al., 2008). A more scalable, routine assay is currently being developed by Zymo Research that combines methyl-specific restriction enzymes and quantitative PCR to determine the percentage of promoter methylation.

1.4   OUTLOOK FOR THE FUTURE OF iPSC BIOBANKING

The field of iPSC biobanking is destined to undergo significant growth in the next ten years, as the technologies for reprogramming differentiated somatic cells (extended to more source cell types, for example), as well as for directing iPSC differentiation along defined lineages, are further refined. As with any emerging field, its community will be faced with numerous challenges, among them regulatory, technical, and logistical. It will be of vital importance for funding agencies such as the NIH to collaborate closely with directors of research projects and Institutional Review Boards (IRBs) to ensure that patient consents are utilized that adequately describe the potential uses of iPSCs in research as well as translational medicine.

The iPSC field is progressing at a remarkable rate, with new techniques published monthly. As major entities such as the NIH, industry, and private foundations begin the process of designing the biological repository collections and research programs of the future, it will be necessary to develop standard techniques and quality control to ensure that results from different studies will be comparable (Chiu and Rao, 2011; Rao and Malik, 2012). Meta analysis of data generated from multiple labs, essential to the elucidation of complex genetic diseases such as schizophrenia, necessitates this level of uniformity. The painful lessons of the early enthusiastic days of the genomics field, when different microarray platforms could not be readily combined to yield useful metadata (Larkin et al., 2005), must be heeded. As of the writing of this book, the authors favor the use of the Sendai reprogramming method for considerations of efficiency, versatility in terms of source cell reprogramming efficacy, and reproducibility. Quality control assays such as gene expression analysis of markers for pluripotency and evaluation of genomic stability by a combination of cellular assays (G banded karyotyping) and array CGH should be considered prior to widespread distribution of iPSCs. While in vivo assays of pluripotency such as the teratoma assay were used in the early days of the field, a consensus is emerging to move away from these expensive and laborious assays and adopt in vitro assays such as embryoid body formation.

These agencies will also be faced with decisions regarding the choice of tissue as source material for reprogramming. Fibroblast cells cultured from skin punch biopsies have been used by many groups who developed the earliest protocols for iPSC reprogramming (for example, see Park et al., 2008). Fibroblasts have the advantage of being relatively easy to grow in culture and are generally amenable to reprogramming by many of the methods in use, including retroviral, lentiviral, episomal, and Sendai vectors. However, participant recruitment is complicated by the fact that people often express reservations about consenting to a punch biopsy, fearing pain, scarring, and the inconvenience of having to be referred to a clinician for the collection. On the other hand, blood can be drawn by any of a large number of highly skilled phlebotomists in a procedure that many people are already comfortable with, and for which scarring is not a concern. One very important advantage of using CPLs or LCLs as source cells for iPSC reprogramming concerns the availability of large well characterized collections such as that of the NIMH Center. Collections such as these are the result of years or decades of work in patient ascertainment and diagnosis, biospecimen processing using highly standardized protocols detailed in SOPs, and careful oversight to ensure that the specimens are as disseminated as widely as possible to the research community. When combined, the development of techniques for converting lymphocytes to iPSCs, described in this book, with the availability of such large centralized collections, have made it possible to contemplate the undertaking of studies of iPSCs and cells differentiated from them into cell types of relevance to psychiatric disorders, using currently available materials that are linked with rich and detailed clinical information collected over a period of years.

Looking more broadly towards the future, we predict that iPSC may supplant LCLs as the main instrument for driving both genetic and neuroscience research. For years, the immortalization of lymphocytes into EBV-transformed cell lines produced a robust source of genomic DNA that was relatively inexpensive to produce and suitable for most genomic analyses. As we transition to genome-level analysis, there is concern that LCLs upregulate telomerase, leading to chromosomal variations, after many cell divisions (Okubo et al., 2001). While iPSC have reported to exhibit genomic variations as well (Gore et al., 2011; Hussein et al., 2011), recent studies suggest that this may be due to genomic variation in source cells more than the iPSC reprogramming itself (Nazor et al., 2012). If this key issue of genomic variance can be controlled or if it turns out to be minimal, iPSC may become a better immortalized cell source than LCLs. Of course, having iPSC provides much more biological power for mechanistic studies of disease or therapy screening. Large numbers of iPSC from a repository would provide a culture-based source of normal human variation that is lacking in studies of drug efficacy before clinical trials. At this point, only the high cost of preparing iPSC argues against a complete transformation of biorepository strategies. We have begun to scale down production of iPSC from a more accessible and well-tolerated tissue donation (blood). Repository processes are already optimized for the receipt of blood samples so this provides a nice transition. We believe that it will be possible to create iPSC from a single drop of blood, such as from a finger stick, making collection much more acceptable and inexpensive. As we perfect the process, scale down even smaller, and increase the efficiency of producing and caring for iPSC, we predict that iPSC repositories will become a widely used and powerful tool for solving the problems of mental health disorders.

CHAPTER 2

Appendix

2.1 MOUSE EMBRYONIC FIBROBLAST (mEF) PREPARATION

Purpose: Isolating, expanding irradiating and freezing mouse embryonic fibroblasts (mEFs).

Materials:

1. CF–1™ mice (Charles River, Strain code 023)
2. 70% Ethanol
3. Sterile dissecting scissors
4. Gelatin-coated flasks
5. PBS
6. Sterile dishes
7. Sterile iris micro-dissection scissors
8. Sterile forceps
9. Trypsin-EDTA
10. mEF Medium
11. mEF Freezing Medium
12. Hemocytometer
13. Planer Kryo 560-16 Controlled Rate Freezer
14. Parafilm
15. Faxitron X-ray Cabinet Irradiator

Procedure:

2.1.1 Harvesting Embryos

a. Sacrifice a mouse at 12.5 day post coitum (dpc).
b. Place mouse abdomen up on a clean paper towel.
c. Thoroughly wet mouse abdomen with 70% ethanol.
d. Use sterile scissors to cut through the peritoneum and into the abdominal cavity to expose uterine horns.
e. Remove uterine horns and transfer to a biosafety cabinet.

2.1.2 Dissecting Embryos to Obtain Fibroblasts

a. Prepare gelatin-coated T75 flasks according to "Preparation of Gelatin-Coated Plates" protocol.

b. Wash uterine horns three times with 10 ml of PBS and transfer to a sterile Petri dish.

c. Under the dissecting scope, cut the uterus into sections each containing 1 embryo.

d. Remove the embryos from the uterine tissue by inserting the forceps into end of each section of the uterus and tear the uterine muscle to release the embryo within the yolk sac and the placenta.

e. Separate the embryo from the yolk sac and placenta.

f. Wash the embryos three times with PBS.

g. Remove the head from each embryo.

h. Using the micro-dissection scissors and forceps remove the heart, lungs and liver from each embryo.

i. Transfer the embryos to a new dish and wash three times with PBS.

j. Mince embryos with the micro-dissection scissors until the pieces are about 1 mm^2.

k. Add 5 ml of trypsin-EDTA to the dish of minced embryos.

l. Incubate at 37°C for 10 minutes.

m. Add 5 more ml of Trypsin-EDTA to the dish.

n. Pipet up and down to continue digesting the tissue.

o. Incubate at 37°C for 20 minutes.

p. Pipet up and down vigorously until the suspension is uniform with few large chunks of tissue (if necessary incubate at 37°C for 10 more minutes).

q. Add 20 ml of mEF medium to the mixture.

2.1.3 Fibroblast Expansion

a. Divide the cell suspension evenly between T75 flasks using 1 flask for every 3 embryos processed.

b. Add mEF medium so that the final volume in each flask is 25 ml.

c. Incubate at 37°C in a 5% CO_2 incubator until flasks are 90% confluent.

d. Passage with trypsin EDTA (this will be passage 2 and the cells should be frozen at this point).

    i. Rinse flasks twice with PBS.

    ii. Add enough trypsin-EDTA to cover bottom of the flask.

    iii. Incubate at 37°C for 5 minutes.

    iv. Add 10 ml of mEF medium.

      v.   Pipet to remove cells from flask.

     vi.   Plate 1:3 in fresh gelatin-coated flasks with <u>mEF medium</u>.

  e. Once cells reach 90% confluence, harvest cells with Trypsin-EDTA.

      i.   Rinse flasks twice with PBS.

     ii.   Add enough trypsin-EDTA to cover bottom of the flask.

    iii.   Incubate at 37°C for 5 minutes.

  f. Count cells following "<u>Hemocytometer Counting</u>" Protocol.

  g. Resuspend mEFs at a final density of 3×10^6 cells per vial in <u>mEF freezing medium</u>.

  h. Freeze with controlled rate freezer and store in liquid N_2.

2.1.4   Expansion of Fibroblasts for Irradiation

  a. Thawing frozen mEF:

      i.   Prepare gelatin-coated T75 flasks according to "<u>Preparation of Gelatin-Coated Plates</u>" protocol.

     ii.   Thaw frozen mEF on the LN_2 vapor phase for 45 minutes.

    iii.   Place mEF vial in 37°C water bath until cells are thawed (~1–2 minutes).

    iv.   Transfer to 9 ml of <u>mEF medium.</u>

     v.   Count cells to determine viability and cell number according to "<u>Hemocytometer Counting</u>" protocol.

    vi.   Centrifuge in table top centrifuge at 1,000 rpm for 5 minutes.

   vii.   Aspirate all gelatin solution from the pre-warmed coated gelatin T75 flask.

  viii.   Resuspend cell pellets in 13 ml of <u>mEF medium</u>, and plate on the warm gelatin T75 flask.

    ix.   Incubate the flasks at 37°C at 5% CO_2/18% O_2.

  b. Passage:

Important—Do not let cells become confluent

Note: Passage two times before irradiating

      i.   Prepare three T75 gelatin-coated flasks according to "<u>Preparation of Gelatin-Coated Plates</u>" protocol.

     ii.   Remove confluent mEF T75 from incubator and check for the percentage confluent:

        1.   If the flask is 75% to 85% confluent proceed to step 3.

        2.   If the flask is less than 75% confluent refresh the medium and place back to incubator until the next day.

    iii. Aspirate medium from flask.

    iv. Rinse flask with PBS.

    v. Add enough trypsin-EDTA to cover the bottom of the flask.

    vi. Incubate at 37°C for 5 minutes.

    vii. Add mEF medium to the flask so that the total volume is about 14 ml.

    viii. Pipet up and down to dislodge cells and transfer to a 15 ml tube.

    ix. Centrifuge in table top centrifuge at 1,000 rpm for 5 minutes.

    x. Resuspend in 9 ml of mEF medium.

    xi. Divide equally between three T75 gelatin-coated flasks and add mEF medium to a final volume of 15 ml.

    xii. Incubate the flasks at 37°C at 5% CO_2/18% O_2.

    xiii. When flasks reach 75–85% confluent repeat steps 2.14.b.i–2.14.b.xii to obtain a total of nine T75 flasks.

2.1.5   Irradiation of Fibroblasts

a. Aspirate medium from flask.

b. Rinse flask with PBS.

c. Add enough trypsin-EDTA to cover the bottom of the flask.

d. Incubate at 37°C for 5 minutes.

e. Add 6 ml of mEF medium to each flask.

f. Pool all of the to one 50 ml centrifuge tube.

g. Pool all three 50 ml centrifuge tubes, and bring the volume up to ~25 ml.

h. Count cells at a dilution of 1:20 (i.e., 50 μl of cells suspension to 950 μl of PBS) following the "Hemocytometer Counting" protocol.

i. Centrifuge in table-top centrifuge at 1000 rpm for 5 minutes.

j. Resuspend at 2×10^6 cells/ml in PBS.

k. Seal tubes with Parafilm.

l. Irradiate at 100 kVp (peak kilovoltage) for 47 minutes (total = 6000 rad).

m. Centrifuge in table top centrifuge at 1000 rpm for 5 minutes.

n. Resuspend in mEF freezing medium at a concentration of 1×10^6 cells/ml.

o. Freeze in 1 ml aliquots.

2.2 PREPARING IRRADIATED mEF PLATES

Purpose: Prepare mouse embryonic fibroblasts (mEFs) to use as feeder layers for human pluripotent stem cells.

Materials:

1. mEF Medium
2. 0.1% Gelatin

Procedure:

1. Coat dishes with gelatin following "Preparation of Gelatin-Coated Plates" Protocol.
2. Thaw frozen irradiated mEFs in LN_2 vapor phase for 45 minutes.
3. Place irradiated mEF vial in a 37°C water bath until cells are thawed (~1–2 minutes).
4. Transfer thawed cells into a 15–ml tube containing 9 ml of mEF medium.
5. Count cells using "Hemocytometer Counting" Protocol.
6. Centrifuge in table top centrifuge at 1,000 rpm for 5 minutes.
7. Resuspend cell pellet in 10 ml of mEF medium, and plate at a density of 10,000–12,000 cells/cm^2 on gelatin-coated dishes.
8. Incubate the mEF dishes at 37°C at 5% CO_2/18% O_2.
9. Cultures are best if used 1–2 days after plating.

2.3 FIBROBLAST CONVERSION

Purpose: isolate, expand and infect fibroblasts with the reprogramming factors hOct4, hSox2, hKlf4, and hc-Myc to generate induced pluripotent stem cells.

Materials:

1. 4% Lidocaine
2. Tegaderm dressing
3. Small screw top specimen container (cryovial) containing sterile culture medium
4. 3–4 mm skin punch
5. Sterile gauze and alcohol or alcohol prep pads
6. Human fibroblast medium
7. Autoclaved forceps
8. Autoclaved scissors
9. Human Fibroblast Enzyme Digestion Medium
10. Scalpel

11. PBS
12. HBSS
13. <u>LB Broth</u>
14. Qiagen Midi- or maxi-prep kit
15. Ultra Clean Endotoxin Removal Kit (MoBio, Calatog number 12615)
16. Plat A cells (Cell Biolabs, catalog number RV-102)
17. <u>Plat A medium</u>
18. <u>Plat A freezing medium</u>
19. Trypsin-EDTA
20. Mr. Frosty Freezing Container (Fisher Scientific, catalog number 15-350-50)
21. Fugene 6 (Roche Applied Science, catalog number 11814443001)
22. 0.45 µm PVDF Filters
23. Polybrene (Sigma, catalog number 107689)
24. pMXs-hOCt3/4 (Addgene, catalog number 17217)
25. pMXs-hKLF4 (Addgene, catalog number 17219)
26. pMXs-hSox2 (Addgene, catalog number 17218)
27. pMXs-hc-Myc (Addgene, catalog number 17220)
28. pMXs-IRES-GFP (Cell Biolabs, catalog numberRTV-013)
29. <u>KOSR</u> or <u>mTeSR$^®$1</u>

Procedure:

2.3.1 Skin Biopsy Collection

Note: Thaw or equilibrate the collection vial containing culture medium for 1 hour at room temperature prior to collection. Do NOT store the collected specimen in the refrigerator or freezer prior to cell culture processing: room temperature only.

a. Written informed consent should be obtained from the subject or the subject's parents for participation in the research study. The risks of participation are reviewed. It is discussed that no direct benefit to study subject is anticipated. It should be noted that the risks of skin biopsy include, but are not limited to, bleeding and infection. It is expected that a scar will form at the biopsy site.

b. A pea sized amount of LMX (4% lidocaine) is applied to the biopsy site (location selected by health care professional performing the biopsy) for 25 minutes under occlusion with a Tegaderm dressing. The biopsy site is then prepared in a sterile manner.

Note: The skin for the biopsy should be prepped using alcohol. If the skin has been prepped using Betadine or chlorhexidine please remove as much as possible of these materials using alcohol. Betadine or chlorhexidine may impede the growth of skin cells in culture.

    c. Local anesthesia may be augmented using 1% lidocaine injected subcutaneously. The biopsy site is cleaned again using multiple alcohol swabs. Numbing of the skin is confirmed by the subject's lack of reaction to a needle prick at the prepped site. A 3-mm punch is used to obtain a biopsy sample in a sterile manner. Pressure is held on the wound until no oozing is observed and a sterile dressing is applied with a small amount of bacitracin ointment. Verbal and written instructions for care of the biopsy site should be provided to the subject and/or the subject's family. Wound care supplies should also be provided.

    d. If the skin sample cannot be processed for culture immediately, it is placed in a small screw top container (autoclaved labeled cryovial completely filled with pre-aliquoted <u>human fibroblast medium</u>. If the container is not completely filled the specimen could possibly stick to the lid and dry out, particularly if it is stored overnight or shipped to a central processing laboratory. These prefilled vials may be stored at 4°C for up to one week. If they will not be used within that period of time, they should be kept at −20°C for up to one month.

Note: The tube will be filled to very close to the top. Take great care when placing the tissue in the tube and closing the lid to avoid contact with the outside of the tube, or spilling the medium, as this may increase the risk of contaminating the sample.

2.3.2 Fibroblast Isolation

    a. Processing the tissue biopsy for cell culture—Day 1:

        i. Remove parafilm wrap from the cryovial containing the skin biopsy and culture medium. Use an alcohol swab to aseptically clean around the cryovial cap, and loosen the cap.

        ii. Carefully remove half of the liquid in the cryovial, and slowly dispense in a well of a multiwell tissue culture plate.

        iii. To determine the culture plate for the enzymatic digestion of the skin biopsy, follow these Tissue Size Guidelines:

           1. Tissue <3 mm: 24 well plate

           2. Tissue 3–5 mm: 12 well plate

           3. Tissue >5 mm: 6 well plate

iv. With a 2–ml pipette remove the rest of liquid in the tube, including the tissue, and slowly transfer to another well.

v.   Inspect both wells and the cryovial for tissue which may carry over in the medium or have remained in the cryovial.

vi. Use an autoclaved forceps to transfer the tissue to a clean (unused) well and add human fibroblast enzyme digestion medium to cover the tissue completely, according to the table below:

1. For a 24 well plate: use 0.5 ml of human fibroblast enzyme digestion medium.
2. For a 12 well plate: use 1 ml of human fibroblast enzyme digestion medium.
3. For a 6 well plate: use 1.5 ml of human fibroblast enzyme digestion medium.

vii. Incubate the tissue sample in human fibroblast enzyme digestion medium 16–18 hours at 37°C in a 5% CO_2/18% O_2 incubator. For biopsies smaller than 3 mm in diameter, incubate in human fibroblast enzyme digestion medium for 4 hours in a 5% CO_2/18% O_2 incubation.

Note: Do not digest longer than 20 hours.

b. Processing the tissue biopsy for cell culture—Day 2:

i. Setting up the explant culture:

1. Using a clean scalpel, score an X onto two clean wells in a multiwell dish, taking care not to puncture the plastic plate. Label the second scored well as "Explant."
2. Use an autoclaved forceps to remove the tissue from the digested well and place it on the first scored well.
3. Remove epidermal layer from the sample by dissection and discard in disposal container.

Note: This step can be omitted if the tissue is less than 3 mm in size.

4. Place the remaining undigested part of the tissue (dermal layer) on the center of the second scored X well plate. Allow it to dry for 2–5 minutes in the hood to allow the tissue fragment to adhere to the dish.
5. Slowly add human fibroblast medium and make sure the fragment stays attached to the dish surface:
   a. 0.5 ml human fibroblast medium for 24-well plate
   b. 1 ml human fibroblast medium for 12-well plate
   c. 2.5 ml human fibroblast medium for 6-well plate
6. Place the explant plate in a 37°C in a 5% CO_2/18% O_2 incubator. Do not move or disturb the culture for 3 days.

ii. Setting up the supernatant culture:
   1. While waiting for the explant tissue to adhere to the well, collect the remaining Digestion medium from the overnight (or 4 hours) digestion well with a pipet, and transfer to a 15-ml centrifuge tube.
   2. Rinse the well twice with 2 ml warm (37°C) PBS, collect and combine in 15 ml centrifuge tube with collected digestion medium.
   3. Bring the final volume in the 15-ml centrifuge tube to 10 ml with PBS. Gently pipet up and down two to three times to dissociate clumps, centrifuge at 1,000 rpm for 5 minutes. Discard supernatant.
   4. Resuspend cell pellet in human fibroblast medium and place in a well labeled as "supernatant." Guidelines for resuspension of the pellet:
       a. 0.5-ml human fibroblast medium for 24-well plate
       b. 1 ml human fibroblast medium for 12-well plate
       c. 2.5 ml human fibroblast medium for 6-well plate
   5. Both explant and supernatant wells should remain undisturbed in a 37°C 5% CO_2/18% O_2 incubator for 3 days before the medium is changed.
   6. After the third day, visually inspect all cultures daily for any signs of microbial contamination.
   7. One week after culture initiation, penicillin/streptomycin can be omitted from culture medium. Depending on the age of the donor, and the collection of the biopsy, cells with fibroblast morphology begin to be detectable within 2–3 days of culture.

2.3.3   Fibroblast Expansion

a. After cells cover two thirds of the well or form confluent foci at several places with well, passage to next two empty wells in the multiwell dish.
b. Freeze cells at third or fourth passage.

Important Note: Passage cells when they achieve 70% confluence.

c. If cultures grow out to 70% confluence in 3–4 days after the first passage, the passaging schedule in 2.3.3.c.i can be used. If cultures take 6–7 days or longer to achieve 70% confluence, use the schedule in 2.3.3.c.ii or 2.3.3.c.iii.
   i. Passaging schedule for "fast growing1" cultures (1:9):
       1. Passage 1: 1 well in 24 well plate → 2 wells in 6 well plate (Final cell suspension volume = 2.5 ml/ well)

2. Passage 2: 1 well in 6 well plate → 3 × T25 flask→ total of 6 T25 flasks will be created/cell line.
3. Passage 3: 1 × T25 flask → 3 × T75 flask → Total of 18 × T75 flasks will be created/cell line.

ii. Passaging schedule for "slow growing2" cultures (1:9):
1. Passage 1: 1 well in 24 well plate → 2 wells in 6 well plate (Final cell suspension volume = 1.5 ml/ well).
2. Passage 2: 1 well in 6 well plate → 3 × T25 flask → total of 6 T25 flasks will be created/cell line.
3. Passage 3: 1 ×T25 flask → 3 × T75 flask → Total of 18 × T75 flasks will be created/cell line.

iii. If at 4 days post-digestion, there is evidence of cell attachment in <10% of well, use the passaging schedule for "poor growing3" cultures (1:3):
1. Passage 1: 1 well in 24 well plate → 1 well in 6 well plate → (final cell suspension volume = 1.5 ml).
2. Passage 2: 1 well in 6 well plate → 1 × T25 flask → total of 1 T25 flask will be created/cell line.
3. Passage 3: 1 × T25 flask → 3 × T75 flask → Total of 3 × T75 flasks will be created/cell line.
4. Cryopreserve 1 × T75 flask at passage 3 and continue expand 2 × T75 flasks at Passage 4.
5. Passage 4: 1 × T75 flask → 3 × T75 flask → Total of 6 × T75 flasks.

d. Procedure for passaging cells (applies to all passages, volumes will vary depending on the starting and ending container types):
i. Equilibration of medium in recipient containers (not necessary for 24- and 6-well plates) (day before passage): 24 hrs prior to performing this procedure prepare recipient containers by dispensing human fibroblast medium into each container to be split as follows, and incubate overnight at 37°C 5% CO_2/18% O_2 incubator.

ii. Trypsinization and Plating step (day of passage):
1. Reserve 1 ml of human fibroblast medium for mycoplasma testing.

CONTAINER TO SPLIT	RECIPIENT FLASK AND SPLIT			
	6 well (1:6)	T25 (1:9)	T75 (1:9)	T175 (1:9)
24 well plate	N/A	3.5 ml	N/A	N/A
6 well plate	1.5 ml	3 ml	7 ml	N/A
T25	N/A	4 ml	9 ml	N/A
T75	N/A	N/A	12 ml	22 ml

2. Tilt the flask, remove and discard all of the remaining human fibroblast medium from the bottom corner of the flasks.

3. Rinse by adding HBSS to each flask and gently rock flasks back and forth (about ten times) to rinse the adherent cells, until you notice an orange-salmon color change.

CONTAINER	VOLUME HBSS/CONTAINER
24 well plate	0.5 ml
6 well plate	2 ml
T25	7 ml
T75	10 ml

4. Tilt the flask and remove all of the HBSS and discard.

5. Dispense trypsin to the container:

CONTAINER	VOLUME TRYPSIN/CONTAINER
24 well plate	0.5 ml
6 well plate	1 ml
T25	1 ml
T75	3 ml

6. Tighten the cap on the flask and place in a 37°C incubator for 4 minutes.

7. After the incubation period, remove flasks from incubator, keeping caps tightened, and gently tap the flask on a hard surface twice to dislodge any attached cells. Check under a microscope that adherent cells have detached effectively.

8. Slowly add <u>human fibroblast medium</u> to inactivate the trypsin, allowing the medium to drop onto the tissue culture surface just inside of the neck while gently rocking side to side so the medium washes any residual cells off the flask.

CONTAINER	VOLUME HUMAN FIBROBLAST MEDIUM/CONTAINER
24 well plate	1–2 ml
6 well plate	2–3 ml
T25	11 ml
T75	12 ml

*Volumes are dependent on the split ratio

9. Gently pipette up and down two times to break any cell clumps pooled in the bottom of the flask.

10. To plate: add appropriate volume of cell suspension to pre-labeled recipient containers according to the table below.

11. Rock gently to cover the surface, loosen caps on all flasks, and place in a 37°C 5% CO_2/18% O_2 incubator for 24 hours.

CONTAINER TO SPLIT	RECIPIENT CONTAINER AND SPLIT (VOLUME OF CELL SUSPENSION PER FLASK/WELL)					
	6 well (1 well) (1:3)	6 well (2 wells) (1:6)	T25 (1:3)	T25 (1:6)	T75 (1:3)	T75 (1:6)
24 well plate (per well)	1.5 ml	2.5 ml	N/A	N/A	N/A	N/A
6 well plate (per well)	N/A	N/A	3 ml	1.5 ml	1 ml	N/A
T25	N/A	N/A	4 ml	2 ml	12 ml	6 ml
T75	N/A	N/A	N/A	N/A	5 ml	2.5 ml

c. Medium exchange (day after passage)

    i. 24 hours after splitting: Warm <u>human fibroblast medium</u> in a 37°C water bath for a minimum of 20 minutes as follows:

CONTAINER	VOLUME HFIB MEDIUM/CONTAINER
6 well plate	1.5 ml → 2.5 ml*
T25	5 ml
T75	13 ml

*Depends on cell growth status

    ii. Remove the newly split flasks from the incubator and tighten caps.

    iii. Working with one flask at a time, move a flask into the bio-safety cabinet and loosen cap.

    iv. Aspirate and discard all of the media in the fibroblast culture and replace with a volume of <u>human fibroblast medium</u> as indicated in the table above.

    v. Incubate at 37°C, 85% RH, 5% CO_2 until adherent cells become confluent on flask surface.

d. <u>Testing Fibroblast Cultures for Mycoplasma Contamination</u>

    i. When cultures have reached 70% confluence at passage 1, aspirate medium from the culture, reserving 1 ml in a prelabeled cryotube for PCR-based mycoplasma testing.

2.3.4 Fibroblast Freezing

Fibroblast cultures are harvested for cryopreservation when they reach ~70% confluence.

a. Harvesting cells for cell counting (working with three T75 flasks at a time):

    i. Aspirate and discard all culture medium.

    ii. Dispense 10 ml HBSS per flask and rock gently. Aspirate and discard the HBSS.

    iii. Dispense 3 ml of trypsin to each flask and rock gently to evenly coat the flask surface. Incubate for 4 minutes at 37°C. Do not exceed 4 minutes.

    iv. Add 12 ml <u>human fibroblast medium</u> to each flask, pipet up and down to break up clumps, and pool the cells from all three flasks in a single 50-ml centrifuge tube.

    v. Repeat steps i–iv until all flasks have been trypsinized.

    vi. Centrifuge tubes for 8 minutes at 1,000 rpm.

    vii. Discard supernatant, and resuspend all pellets in 40 ml (Vol Dilution) of <u>human fibroblast medium</u>.

b. Counting cells using the Vi-Cell:
   i. Add 0.75 ml of <u>human fibroblast medium</u> to Vi-Cell sample cup.
   ii. Gently mix the cells suspension in a 50-ml centrifuge tube, remove 0.25 ml and put into Vi-Cell sample cup.
   iii. Run sample on Vi-Cell per manufacturer's operating instructions, using the following guidelines:
      1. Select "Fibroblast" cell type.
      2. Parameter settings:

Minimum diameter (Microns)	10
Maximum diameter (Microns)	30
Number of Images	50
Aspirate cycles	1
Trypan blue mixing cycles	3
Cell brightness (%)	85
Cell sharpness	100
Viable cell spot brightness (%)	65
Viable cell spot area (%)	5
Minimum circularity	0
De-cluster degree	Medium

      3. Set Dilution Factor to 4. Run samples.
      4. Cells are cryopreserved at 1.5×10^6 cells/1 ml per ampule for master stock storage. The volume of <u>human fibroblast freeze medium</u> for cell pellet resuspension is calculated according to the formula:

$$\frac{\left(\text{Viable} \dfrac{\text{Cells}}{\text{ml}}\right) \times \text{Volume of dilution 40 ml}}{1.5 \times 10^6 \text{cells}} = \text{Volume of Freeze media (ml)}$$

      5. Centrifuge tubes for 8 minutes at 1,000 rpm. Aspirate and discard supernatant.
      6. Dilute cell suspension with X ml (calculated in step 4 above) $1 \times$ <u>human fibroblast freeze medium</u> dropwise to cell suspension and aliquot 1 ml per cryovial/ampule.

7. Sealing ampules for storage of master stocks: to prepare master stocks of cryo-preserved fibroblasts, prepare enough cells for 20–30 glass ampules, which are then sealed using a flame sealer for storage in liquid nitrogen.

8. Cryopreservation: freeze fibroblast cells in glass ampules for permanent storage in liquid nitrogen tanks by gradual freezing using a Forma Control Rate Freezer (operated as per manufacturer's instructions).

9. Permanent storage of frozen fibroblasts: store sample tubes in assigned locations in LN_2 tanks.

2.3.5   Plasmid Prep

a. Inoculate 5 ml of <u>LB broth</u> with a single bacterial colony.

b. Grow at 37°C with shaking for ~8 hours.

c. Transfer 5 ml of bacterial culture to 50–250 ml of <u>LB broth</u> (note: amount of overnight bacterial culture to prepare depends on whether you plan to do a midi or maxiprep of the DNA and which method you use to obtain and purify the DNA).

d. Grow overnight at 37°C with shaking.

e. Follow the instructions of the desired method for obtaining plasmid DNA (note: retro-viral production is better when a kit that minimizes endotoxin presence is used such as a Qiagen Maxi or Midi Prep kit followed by the UltraClean Endotoxin Removal Kit from MoBio).

2.3.6   Plat A Cell Culture

a. Thawing:
   i.   Place vial in 37°C water bath until cells are thawed.
   ii. Transfer to 15 ml tube containing 9 ml of <u>Plat A medium</u>.
   iii. Centrifuge in table top centrifuge at 1,000 rpm for 5 minutes.
   iv. Resuspend in <u>Plat A medium</u> without antibiotics.
   v.   Plate 18,000 cells/cm^2.
   vi. Incubate at 37°C at 5% CO_2/18% O_2.

b. Passaging Plat A Cells.

Note: Do not let cells become confluent
   i.   Rinse plate with PBS.
   ii. Add enough Tryspin-EDTA to cover the well.
   iii. Incubate at 37°C for 5 minutes.
   iv. Transfer cell suspension to a 15–ml tube containing <u>Plat A medium</u>.

     v.   Centrifuge in table top centrifuge at 1,000 rpm for 5 minutes.

     vi. Resuspend in <u>Plat A medium</u>.

     vii. Plate on new dishes at a density of 2–5 × 10^4 cells/cm^2.

  c.   Freezing Plat A Cells:

     i.   Rinse plate with PBS.

     ii. Add enough Tryspin-EDTA to cover the well.

     iii. Incubate at 37°C for 5 minutes.

     iv. Transfer cell suspension to a 15–ml tube containing <u>Plat A medium</u>.

     v.   Centrifuge in table top centrifuge at 1,000 rpm for 5 minutes.

     vi. Count cell number following "Hemocytometer Counting" protocol.

     vii. Resuspend 1 to 2 × 10^6 cells in 1–ml <u>Plat A freezing medium</u>.

     viii. Freeze in Mr. Frosty Freezing Container at −80°C overnight.

     ix. Transfer to LN$_2$.

2.3.7 Transfection of Plat A cells

  a.   Day 1:

     i.   Seed Plat A cells at 4 × 10^6 cells per 10 cm dish.

  b.   Day 2:

     i.   Change medium on Plat A cells to <u>Plat A medium</u> without antibiotics.

     ii. For each 10 cm dish, transfer 0.3 ml of antibiotic medium to an microfuge tube.

     iii. To each microfuge tube add 27 µl of Fugene.

     iv. Mix with finger tapping and incubate at room temperature for 5 minutes.

     v.   Add 9 µg of each plasmid DNA to each microfuge tube.

     vi. Mix by finger tapping and incubate at RT for 15 minutes.

     vii. Add the DNA/Fugene solution to the 10 cm dish and incubate overnight at 37°C.

  c.   Day 4:

     i.   Collect medium off of 10 cm retrovirus production dishes.

     ii. Filter with 0.45 µm filter.

2.3.8 Fibroblast Infection

  a.   Plating Fibroblasts (Day 1):

     i.   Rinse plate with PBS.

     ii. Add enough Tryspin-EDTA to cover the fibroblasts.

     iii. Incubate at 37°C for 5 minutes.

     iv. Transfer cell suspension to a 15 ml tube containing <u>fibroblast medium</u>.

     v.   Count cells following "<u>Hemocytometer Counting</u>" Protocol.

    vi. Centrifuge in table top centrifuge at 1,000 rpm for 5 minutes.

    vii. Remove supernatant and resuspend cell pellet in <u>fibroblast medium</u> at a density of 8.6×10^4 cells/ml.

    viii. Plate 1 ml of cells suspension in 1 well of a 6 well plate.

    ix. Incubate at 37°C at 5% CO_2/18% O_2.

  b. Infecting Fibroblasts with retrovirus (Day 2):

Note: Retroviruses must be collected on the day of infection so timing is important

    i. Prepare virus mix for infection.

Note: you can do a GFP infection control to determine viral transduction efficiency if desired.

      1. Add 0.250 ml of each virus(OSKM).

      2. Add 1.3 ml of <u>fibroblast medium</u>.

      3. Add 1 µl polybrene.

    ii. Aspirate medium from fibroblast dishes.

    iii. Add virus mix medium.

  c. Feeding fibroblasts (Days 3–6):

    i. Change <u>fibroblast medium</u> every other day starting on Day 3.

  d. Passaging to mEFS (Day 7):

    i. Rinse plate with PBS.

    ii. Add enough Tryspin-EDTA to cover the well.

    iii. Incubate at 37°C for 5 minutes.

    iv. Transfer cell suspension to a 15 ml tube containing <u>fibroblast medium</u>.

    v. Centrifuge in table top centrifuge at 1,000 rpm for 5 minutes.

    vi. Prepare mEF dishes according to "<u>preparing irradiated mEF dishes</u>" Protocol.

    vii. Plate on mEFs in <u>fibroblast medium</u> at the following densities:

      1. At 2.5×10^5 cells/cm^2

      2. At 5×10^5 cells/cm^2(this density usually leads to the most colonies)

      3. 7.5×10^5 cells/cm^2

2.3.9   Colony Formation

  a. Monitor for Colony Formation (Days 8–28):

    i. Refresh medium with <u>KOSR</u> or <u>mTeSR®1</u> everyday.

    ii. Evaluate plates for colony formation.

Note: colonies may start to appear around day 16 but it is better not to pick the colonies until after day 21. An example of an early colony is shown in Figure 2.

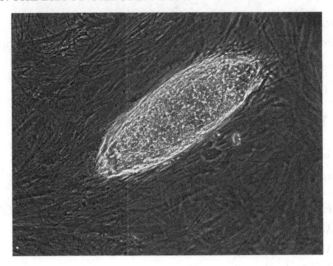

FIGURE 2: Reprogrammed Fibroblasts. An example of the morphology of a colony obtained from reprogrammed fibroblasts at day 15.

2.4 PBMC CONVERSION

Purpose: Isolation of human CD4$^+$ T cells from peripheral blood mononuclear cells (PBMC) by depleting B cells, NK cells, monocytes, platelets, dendritic cells, CD8$^+$ T cells, granulocytes and erythrocytes. Isolated CD4$^+$ T cells are bead- and antibody-free and are suitable for any downstream application. Before Sendai virus infection, these cells must be activated so that they expand and are actively proliferating.

Materials:
1. Whole blood
2. Nycoprep™
3. <u>CPL RPMI</u>
4. <u>CPL freeze medium</u>
5. Invitrogen Dynabead® Untouched™ Human CD4 T cell kits (Life Technologies, catalog number 113.46D) contains:
   a. Antibody mix for human CD4 T cells (a mixture mouse IgG antibodies against the non-CD4$^+$ T cells)
   b. Depletion MyOne™ Dynabeads®
6. <u>CPL isolation/activation buffer</u> (keep on ice before and after each use)

7. DynaMag™-2 magnet (Life Technologies, catalog number 12321 D)
8. Ice cold <u>heat-inactivated FBS</u>
9. <u>Irradiated mEF layer</u>
10. CD3/CD28 Dynabeads® (Life Technologies, catalog number 111.31D)
11. IL-2 (250 U/mL) (Peprotech, catalog number AF-200-02)
12. Cytotune™-iPS Sendai Reprogramming Kit (Life Technologies catalog number, A1378001)
13. PBS
14. 0.1 % Gelatin
15. <u>mEF Medium</u>
16. <u>KOSR</u>
17. Mr. Frosty Freezing Container (Fisher Scientific, catalog number 15-350-50)

Procedure:

2.4.1 PBMC Isolation

a. Preparation:
    i.   Pre-label one empty 15 ml centrifuge tube and one 15-ml centrifuge tube containing 5 ml Nycoprep™ with the appropriate sample ID and barcode labels for each sample to be processed.
    ii. Each blood sample requires two pre-labeled cryogenic preservation vials and pre-assigned locations for liquid nitrogen storage.
b. Centrifugation to produce buffy-coat:
    i.   Centrifuge the vacutainer(s) for 10 min at 3,000 rpm with the brake on high.
    ii. Loosen the caps on the two centrifuge tubes and use an alcohol swab to aseptically remove the stopper of the blood tube.

Notes: Contain and clean any drips or spills immediately before proceeding. Be careful while handling the centrifuged blood in order to prevent mixing of the layers.

c. Layering of buffy coat onto gradient:
    i.   Using a sterile 5-ml pipette remove the buffy-coat layer (i.e., white blood cell layer) at the plasma/RBC interface and place in an empty 15-ml centrifuge tube (Figure 3).

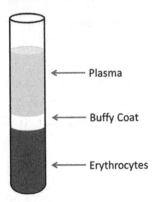

FIGURE 3: Separated Whole Blood. After centrifugation whole blood is separated into 3 layers—plasma, buffy coat and erythrocytes. The buffy coat contains the desired lymphocytes.

*Note: * The buffy-coat should be skimmed off the top of the erythrocyte (red blood cell, "RBC") layer taking as few RBCs as possible. Take as little volume as possible to effectively remove the buffy-coat. This volume may vary between 1 and 3 ml.*

    ii. Dispense enough <u>CPL RPMI medium</u>, warmed to 37°C, to make a final volume of 6 ml in the centrifuge tube containing the transferred cells.

    iii. Mix and gently layer the 6 ml of the blood/media solution on top of the pre-dispensed 5 ml of Nycoprep™.

    iv. Replace caps on the two 15-ml centrifuge tubes.

    v. Dispose of mixing tube in a biohazard trash container.

    vi. Once all the blood samples have been layered, centrifuge for 30 minutes at 2,500 rpm with the brake turned OFF.

  d. Washing the PBMC layer:

    i. First Wash:

      1. Carefully remove the samples from the centrifuge taking care not to disturb the gradients.

      2. Inspect each gradient for separation quality.

      3. Using a 5 ml pipette, remove the lymphocyte layer (white blood cells) at the Nycoprep™/Plasma interface and dispense it into the labeled 15 ml centrifuge tube (Figure 4).

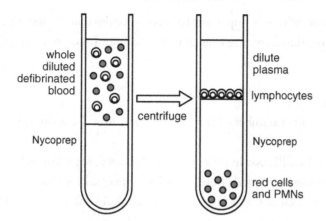

FIGURE 4: Nycoprep™ Separation of Lymphocytes. After Nycoprep centrifugation the desired lymphocytes are present at the plasma/Nycoprep™ interface.

4. Add sufficient <u>CPL RPMI medium</u> to bring the final volume to 10-ml.
5. Once all the layered products have been collected, place the labeled centrifuge tubes in the centrifuge and spin for 10 minutes at 3,000 rpm with the brake turned ON.

ii. Second Wash:
1. Carefully remove the samples from the centrifuge taking care not to disturb the cell pellets.
2. Inspect each cell pellet for separation quality.
3. Using a 10 ml pipette, remove the supernatant.
4. Resuspend the cell pellet in 10 ml <u>CPL RPMI medium</u>
5. Once all of the cell pellets have been re-suspended, place the centrifuge tubes in the centrifuge and spin for 10 minutes at 3,000 rpm with the brake turned ON.
6. Inspect each cell pellet for separation quality.

2.4.2   Cryopreservation of PBMC

a. Using a 10 ml pipette, remove the supernatant.
b. Resuspend the cell pellet in 3 ml of <u>CPL freeze medium</u>.
c. Dispense the freeze media suspended cell pellet solution into two (2) appropriately labeled cryovials.
d. Cryopreservation: PBMCs are frozen using computer controlled rate freezers following the manufacturer's operating instructions. The use of computer controlled rate freezers

for cryopreservation is important because an optimal and constant freezing rate should be maintained to minimize cellular dehydration or ice crystal formation.

2.4.3   CPL Thaw

a. Remove required number of frozen CPL vials from LN_2 storage, and place in 10×10 box.

b. Place 10×10 CPL box in water bath until thawed, approximately 2–3 min.

c. Remove thawed CPL vials from water bath, spray CPLs vials with 70% ethanol.

d. Work with 1 cell line at a time. Using an alcohol swab clean around CPL vial cap and gently loosen the cap.

e. Transfer contents of CPL vial to an appropriate labeled 15-ml tube. Repeat this step for the rest of the CPL vials.

f. Bring volume in each 15-ml tube containing CPLs to 10 ml with CPL RPMI medium and gently mix.

g. Pipette 10 µl cell suspension from each cell line and determine cell counts following the "Hemocytometer Counting" protocol.

2.4.4   CD4+ Cell Isolation and Activation

a. Washing Depletion MyOne Dynabeads:

   i. Vortex Depletion MyOne™ Dynabeads® vial, and gently mix with a 1,000 µl pipette to ensure Dynabeads are homogenous.

   ii. Add 100 µl of Depletion MyOne™ Dynabeads® in 2 ml microfuge tube (1 microfuge tube per cell line).

   iii. Add 1 ml of CPL isolation/activation buffer and mix

   iv. Place tubes on a DynaMag™-2 magnet for 1 minute until Depletion MyOne™ Dynabeads® are attached to the magnet (see the dark brown column line on Figure 5b).

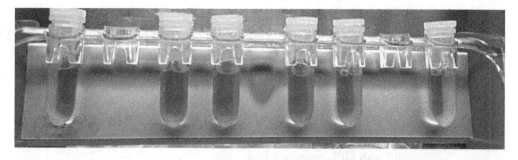

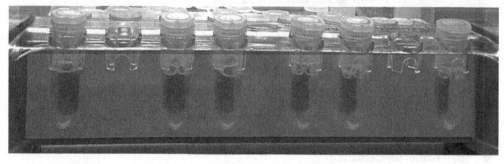

FIGURE 5A AND B: Attachment of Depletion MyOne™ Dynabeads® to DynaMag™-2 magnet. (A) CD4+ negative isolation Depletion MyOne™ Dynabeads® in solution are placed on the DynaMag™-2 magnet. (B) After 1-2 minutes the beads magnetically attracted to the DynaMag™-2 magnet and the supernatant can be removed.

      v.   Discard supernatant.

     vi. Remove tube from magnet and resuspend the washed Depletion MyOne™ Dyna-beads® in 100 μl of <u>CPL Isolation/Activation Buffer</u>.

    vii. Place washed Depletion MyOne™ Dynabeads® tubes in hood until needed.

  b. Thawing CPL:

      i. Remove frozen CPL vials from LN_2 vapor tank.

     ii. Place CPLs in a 37°C water bath until thawed (approximately 2–3 minutes).

    iii. Remove thawed CPL vials from water bath and spray vials with 70% ethanol.

    iv. Remove thawed CPLs and transfer to an appropriate labeled 15 ml tube.

     v. Bring volume in each 15 ml tube containing CPLs to 10 ml with <u>CPL RPMI</u> and gently mix.

     vi. Count cells following "<u>hemocytometer counting</u>" protocol.

    vii. Centrifuge in table top centrifuge at 1500 rpm for 8 minutes.

   viii. Remove supernatant and resuspend CPL pellet at a density of 2×10^6 cells/ml with <u>CPL RPMI</u>.

    ix. For each cell line:

1. Transfer 2×10^6 CPL cells to microfuge tube. This portion will be used for CD4 isolation.
2. The remaining CPL suspension in 15 ml tube will be placed on ice for freezing.

    i. Centrifuge in table top centrifuge at 1,500 rpm for 8 minutes.

    ii. Remove supernatant, resuspend the remaining CPL pellet at 2×10^6 cells/ml with CPL freeze medium.

    iii. Aliquot $2-10^6$ cells per cryovial.

    iv. Transfer to Mr. Frosty Freezing Container and place in −80°C freezer overnight.

    v. Transfer frozen CPL vials in LN_2 vapor tank.

c. Isolating CD4+ T Cells:

    i. Spin 2×10^6 CPL portion in microfuge tubes for CD4 isolation at 3,000 rpm for 5 minutes in microfuge.

    ii. Remove supernatant.

    iii. Resuspend CPL pellet in 100 μl of Isolation/Activation Buffer.

    iv. Add 20 μl of ice cold heat inactivated FBS.

    v. Add 20 μl of Antibody mix for human CD4 T cells and mix well.

    vi. Incubate for 20 minutes at 4°C (in refrigerator—not on ice).

    vii. Wash cells by adding 2 ml of CPL Isolation/Activation Buffer. Mix well by tilting tubes ten times.

    viii. Centrifuge at 3,000 rpm for 5 minutes in microfuge.

    ix. Discard the supernatant.

    x. Resuspend cells in 100 μl of CPL isolation/activation buffer, mix gently.

    xi. Add 100 μl pre-washed Depletion MyOne™ Dynabeads® from step 2.4.4.a.vii. (don't forget to mix the prewashed Depletion MyOne™ Dynabeads® before use).

    xii. Incubate at room temperature for 15 minutes with rotation.

    xiii. Remove tubes from the rotation machine; mix the bead-bound cells by vigorously pipetting more than ten times.

    xiv. Add 1 ml of CPL isolation/activation buffer to each bead-bound cells tubes, mix gently.

    xv. Place the bead-bound cells tubes on the DynaMag™-2 magnet for 2 minutes. Non-CD4⁺ cells will be bound to magnetic column. CD4⁺ cells remain in supernatant.

    xvi. Transfer supernatant which contains CD4⁺ T cells to a new appropriately labeled microfuge tube.

xvii. Repeat steps xiv–xvi with tube containing the Depletion MyOne™ Dynabeads® and combine the two supernatants.

xviii. Gently mix combined supernatants.

xix. Count number of CD4+ cells using "hemocytometer counting" protocol.

xx. Label cell ID number and total CD4+ T cell count for each cell line on a 48-well plate.

xxi. Do not spin CD4+ cell suspension until master mix for CD4+ T cell activation is made.

d. CD4+ Cell Activation:

i. Make sure CD3/CD28 Dynabeads® are completely resuspended by vortexing.

ii. Transfer needed amount of CD3/CD28 Dynabeads® per cells line to a new microfuge tube (see chart).

iii. Add 1 ml of CPL isolation/activation buffer to wash the beads.

iv. Vortex microfuge tube with beads for 5 seconds.

v. Place microfuge tube on DynaMag™-2 magnet for 1 minute to separate beads from of CPL isolation/activation buffer.

vi. Discard supernatant from beads. CD3/CD28 Dynabeads® remain in the microfuge tube and attached to the magnetic column.

vii. Resuspend washed bead with 1 ml of CPL RPMI.

viii. Make master mix for all cell lines based on table below (use 48-well plate protocol for any cell number less than 5×10^5).

1. Combine resuspended washed CD3/CD28 Dynabeads® to a new master mix 15 ml tube with required amount of medium.

2. Add required amount of IL-2 to the master mix tube.

ix. Spin CD4+ cells suspension at 3,000 rpm for 5 minutes in microfuge to remove supernatant.

x. Resuspend each CD4+ tube in required amount of master mix medium, and transfer to an appropriated labeled well (Note: a mixture of CD4+ cells + CD3/CD28 Dynabeads® + IL-2 + CPL RMPI all are in a well).

WELL SIZE	NO. OF CD4+ CELLS	CD3/CD28 DYNABEADS®	IL-2	MEDIUM (15% FBS/ 1% PS/ 1% L-GLUT)
24	1×10^6	24 µl	0.4 µl	1 ml
48	5×10^5	12.5 µl	0.2 µl	0.5 ml

xi. Incubate at 37°C at 5%CO_2/4% O_2—Note that this requires a low-O_2 incubator (Cells before activation are shown in Figure 6A and cells after 2 days of activation are shown in Figure 6A and B).

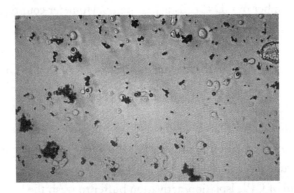

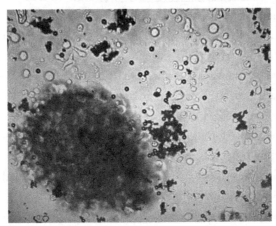

FIGURE 6A AND B: CD3/CD28 Activation of CD4+ T cells. (A) The mixture of CD4+ T cells and CD3/CD28 activation beads immediately after the start of activation. At this time, the dark, spherical beads outnumber the translucent, spherical CD4+ T cells. (B) Forty eight hours after incubation, large clumps of CD4+ T cells become visible.

xii. Feed activated CD4+cells on the third day:
   1. Prepare medium: for every 250 μl of CPL RPMI add 0.1 μl IL-2.
   2. Feed:
      i.   Spin plate in centrifuge at 1,000 RPM for 5 minutes.
      ii. Remove supernatant.
      iii. Resuspend with 250 μl of CPL RPMI +IL-2.
   xiii. Add virus on the fourth day or when CD4+ cells reach up to 50% or more of activated cells as shown by the large clumps of CD4+ T cells in Figure 7.

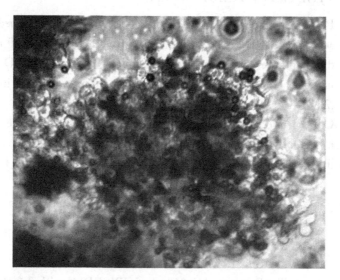

FIGURE 7: CD3/CD28 Activated CD4+ T Cell Ready for Sendai Virus Infection. After four days of stimulation with CD3/CD28 beads, large clumps of CD4+ T cells predominate the culture.

2.4.5   Sendai Infection

a.   (Day −4 to Day 0) Isolate and activate CD4$^+$ T cells.

b.   (Day 0) Sendai Infection:

   i.   Set-up:

      a.   Thaw and aliquot Sendai virus according to "<u>Thawing and aliquoting Sendai Virus</u>" protocol.

      b.   Label 2-ml microfuge tubes corresponding to each CPL cell line.

      c.   Remove Cytotune™-iPS Sendai Reprogramming Kit (hOct4, hSox2, hKlf4, and hc-Myc) from −80°C storage.

         i.   Thaw each tube on ice.

         ii.   Once thawed, briefly centrifuge the tube and place it immediately on ice until use.

      d.   Thaw and seed irradiated mEFs in 4 well plate according to "<u>Preparing mEF Plates</u>" protocol.

   ii.   Separation of Activated CD4$^+$ T cells from CD3/CD28 Dynabeads®:

      a.   Gently mix the activated CD4$^+$ T cells/bead mixture in the 48-well plate by pipetting and transfer to labeled microfuge tube. Note: This mixing will help separate the CD4$^+$ T cells from the activation beads.

      b.   Add required amount of <u>CPL RPMI</u> to pre-labeled microfuge tubes to bring the volume to 1 ml.

    c.   Place microfuge tubes on DynaMag™-2 magnet to removed CD4⁺ cells from activation beads for 1–2 minutes. (Note: CD4⁺ T cells will remain in the medium and the activation beads will remain attached to the DynaMag™-2 magnet (save CD3/CD28 Dynabeads® in case they are needed in step 2.4.5.b.ii.d.ii .2.iv).

    d.   Collect supernatant and transfer to a labeled microfuge tube.

        i.   Gently mix the cells suspension in microfuge tube.

        ii.   Remove 10 µl of cell suspension for counting using "Hemocytometer Cell Counting" protocol.

            1.   If cell count is more than 80,000, go to step 2.4.5.b.ii.e.

            2.   If cell count is less than 80,000:

                i.   Transfer cell suspension to 15 ml tube.

                ii.   Spin down cells at 1,500 rpm for 8 minutes.

                iii.   Resuspend in 250 µl of CPL RPMI with 0.1 µl of IL-2.

                iv.   Mix cell suspension with previously used beads in microfuge tube (from step 2.4.5.b.ii.c.).

                v.   Place beads and cells back into the 24-well plate for additional activation.

                vi.   Incubate and check cell count for the next two days until there are at least 80,000 cells.

    e.   Transfer 80,000 cells to labeled microfuge tubes for infection (cell count/ 80,000 = number of ml to transfer).

    f.   Pellet remaining activated CPLs (in microfuge tube) at 3,000 rpm for 5 minutes.

    g.   Aspirate supernatant and dry cell pellet at −80°C.

        i.   Infection of activated CD4⁺ T cells with Sendai virus.

Note: Each virus (hOct4, hSox2, hKlf4 and hc–Myc) in each lot will have a different titer which can be found on the "Certificate of Analysis" that comes with each Cytotune kit.

    a.   Calculate amount of virus to be used for each infection (Note: each infection should be done at an MOI of 2.5).

        i.   Determine the amount of each virus to be used for each infection. (200,000/viral titer)*1,000 = µL of virus to use.

        ii.   Calculate and fill in chart.

        iii.   For each infection prepare a master mix in a microfuge tube that contains:

            1.   100 µl of CPL RPMI

            2.   Calculated amount of each virus (from above table)

CYTOTUNE VIRUS	HOCT4	HSOX2	HKLF4	HC-MYC
Lot number				
Titer (CIU/ml)				
MOI	2.5	2.5	2.5	2.5
Amount of virus to use for infecting 8×10^4 CD4$^+$ T cells (µl)				

    3.   Thoroughly mix by pipetting gently.

    4.   Complete the next step within 5 minutes.

  iv. Spin down the CD4$^+$ cells from step 2.4.5.b.ii.e. at 3,000 rpm for 5 minutes.

  v. Remove spent medium.

  vi. Resuspend CD4$^+$ T cells in 105 ml of master mix solution which was prepared in step 2.4.5.b.iii.a.iii.

  vii. Transfer to 48-well plate.

  viii. Incubate at 37°C at 5%CO_2/4% O_2 (Low O_2 incubator).

  b. (Day 1) Continue incubating infected cells. Do not replace medium.

Note: You might see some cytotoxicity 24–48 hours post-transduction, which can affect >50% of your cells. This is an indication of high uptake of the virus. We recommend that you continue culturing your cells and proceed with the protocol.

  c. (Day 2) Plate transduced CD4$^+$ T cells on mEF plates prepared in step 2.4.5.b.i.c.

    i. Collect the infected CD4$^+$ T cells.

      1. Transfer to microfuge tube.

      2. Spin plate in microfuge at 3,000 rpm for 5 minutes.

      3. Remove supernatant.

      4. Resuspend with 250 µl of <u>CPL RPMI</u> + 0.1 ml of IL-2.

    ii. Remove <u>mEF medium</u> from 4-well plate containing irradiated mEFs.

iii. Plate the infected CD4$^+$ T cells onto the irradiated mEF plate at a density of 1 infection/well. When examined under a microscope you should see round cells sitting on top of the larger, flat mEFs as seen in Figure 8.

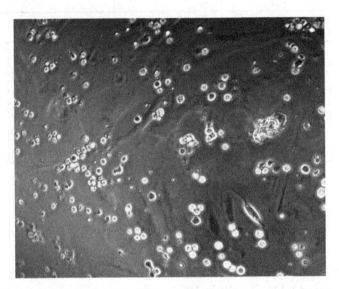

FIGURE 8: Activated CD4+ T cells Infected with Reprogramming Factors. Two days after infection with the Sendai viruses containing the reprogramming factors, the CD4+ T cells are plated on a mEF feeder layer.

iv. Check 4-well plate under the microscope to ensure cells were successfully plated. Infected CD4$^+$ T cells should be spherical and easily distinguished from the large, irregularly shaped, flat mEFS.

v. Incubate at 37°C at 5%CO_2/4% O_2.

d. (Day 3) Change medium to <u>KOSR</u>.

e (Day 4 to 28) Feed and monitor cells.

i. Refresh <u>KOSR</u> every day (200 μl).

ii. Starting on Day 8, observe the plates every other day under a microscope for the emergence of cell clumps indicative of reprogrammed cells. Figure 9A shows a very small colony immediately after appearance and this colony stains positive for alkaline phosphatase as seen in Figure 9B.

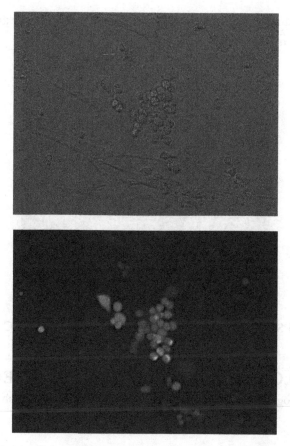

FIGURE 9A AND B: Initial Colonies Generated from Reprogrammed Blood. (A) An example of the morphology of a pluripotent colony found at day 8. (B) The same day 8 colony stained with the Alkaline Phosphatase Live Stain.

iii. Two to three weeks after transduction, colonies should have grown to an appropriate size for transfer as seen in Figure 10.

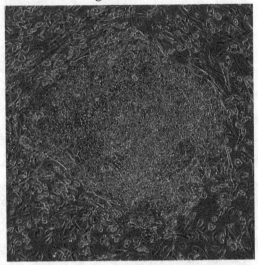

FIGURE 10: Mature iPSC Colony from CPLs. A reprogrammed blood iPSC ready for picking.

2.5   IPSC COLONY IDENTIFICATION AND SELECTION

Purpose: Identify reprogrammed cell colonies and transfer to Matrigel dishes for expansion.

Materials:

1. KOSR
2. mTeSR®1
3. BD Matrigel™ (BD Biosciences, catalog number 354277)
4. Manual Scraping Tool
5. 5 mM Rho Kinase inhibitor Y7632 (Tocris Biosciences, catalog number 1254)

Procedure:

1. During days 14–28 of reprogramming make sure to examine the plates microscopically to monitor for the formation of colonies with pluripotent cell morphology. (Note: these colonies often begin to appear around days 10–14, but it is generally best to wait to manually select these colonies about 1 week after their initial appearance.)
2. Prepare BD Matrigel™-coated plates according to "Preparation of Matrigel Plate" Protocol.

3. Use a Pasteur pipet to make a manual passaging tool as described in the "Manual Scraping Tool" protocol (note: unlike the normal scraping tool, the tool used for selecting reprogrammed colonies should not be flamed on the end to seal the glass into a sphere. Instead, the end should be left broken to provide a sharp edge for cutting).

4. Aspirate medium from the well of reprogrammed cells and replace with KOSR.

5. Focus colony of interest under microscope, cut around colony by using a pasture pipette with sharp end point, and gently pull to detach the colony from the rest of the infected cells on the well. (Note: The colony of interest will be floating in the KOSR medium.)

6. Using 100 µl pipette, transfer a single picked colony to the new well.

7. Add the Rho Kinase Inhibitor Y 2763 to each well at a final concentration of 5 µM.

8. Incubate both the newly picked colony plate and the plate of reprogrammed cells at 37°C at 5%CO_2/4% O_2 (Low O_2 incubator).

9. Refresh the medium daily.

    a. Newly picked colony plate: use KOSR media for transfer of colony and then refresh medium with mTeSR®1 medium until it is ready for passage according to "Dispase Passaging of iPSC" protocol.

    b. Old reprogrammed cell plate: continue feeding with KOSR medium and manually selecting colonies of interest.

2.6   IPSC EXPANSION

Purpose: Expand iPSC colonies for banking from either newly transferred and reprogrammed iPSC or from thawed iPSC.

Materials:

1. BD Matrigel™ (BD Biosciences, catalog number 354277)
2. mTeSR®1
3. Manual Scraping Tool
4. 5 mM Rho Kinase inhibitor Y7632 (Tocris Biosciences, catalog number 1254)
5. 1 U/ml Dispase
6. mFreSR® (Stem Cell Technologies, 05855) (Note: should be kept on ice at all times)
7. Accutase™ (Stem Cell Technologies, Catalog number 07920)
8. PBS

Procedure:

2.6.1 Thawing iPSC

a. Remove the iPSC frozen ampule from liquid N_2 and place it in the N_2 vapor phase for 45 minutes.

b. Prepare a BD Matrigel™-coated 6-well plate according to "Preparation of Matrigel-Coated Plates" protocol.

c. Thaw frozen iPSC in a 37°C water bath for ~1 minute.

d. Transfer cell suspension to a 15-ml tube containing mTeSR®1 medium.

e. Centrifuge in table top centrifuge at 1,000 rpm for 5 minutes.

f. Remove supernatant and resuspend cell pellet in mTeSR®1 medium at a density of 1×106 cells/ml.

g. Transfer 2 µl of cell suspension to the matrigel-coated plate.

h. Add 1 ml of Rho Kinase inhibitor Y7632 to the wells containing thawed iPSC.

i. Gently shake the new plate from left to right and up to down to ensure cells are evenly distributed throughout the well.

j. Incubate at 37°C at 5% CO_2/4% O_2 (Low O_2 incubator).

k. Exchange medium every day with mTeSR®1 until the colonies are 70–80% confluent.

2.6.2 Passaging iPSC

a. Examine colonies daily until they are 70–80% confluent and the colonies are 0.25 to 0.5 mm in diameter (this should take about 7 days).

b. Prepare BD Matrigel™-coated plates according to "Preparation of BD Matrigel™-Coated Plate" protocol.

c. Remove spontaneous differentiation from the well to be passaged according to "Manual Removal of Differentiation" protocol.

d. Dissociate colonies into pieces following "Dispase Dissociation of iPSC" protocol.

e. Plate colony pieces on new BD Matrigel™-coated plate at a ratio of 1:8 to 1:12 depending on cell density and amount of differentiation removed.

f. Incubate at 37°C at 5% CO_2/4% O_2 (Low O_2 incubator).

g. Exchange medium every day with mTeSR®1 until the colonies are 70–80% confluent.

2.6.3 Cryopreservation of an iPSC Single Cell Suspension

a. iPSC cells lines can be frozen if they meet the following criteria:

   i. Expanded to 12–>18 wells of 6 well plates.

   ii. Passage number 4 or greater.

   iii. Colonies look healthy and have low levels of spontaneous differentiation.

b. Scope all wells of the iPSC line that needs to be frozen under microscope to locate the differentiated cells.

c. Follow "<u>Manual Removal of Differentiation</u>" protocol to remove differentiated cells.

d. Aspirate media from the wells.

e. Wash wells 2 × 2 ml of PBS.

f. Add 1 ml of Accutase™ to each well.

g. Incubate at 37°C for 5 minutes.

h. Ensure that the cells are loosened from the plate.

i. Using a 1,000 ml tip, take the cell suspension and rinse wells twice before collecting cells.

j. Transfer cells suspension to 50 ml tube.

k. Add an equal amount of PBS to the 50 ml tube.

l. Remove 10 ml of cell suspension for counting on a hemocytometer.

m. Spin 50 ml tube at 1,000 rpm for 5 minutes.

n. Calculate volume of mFreSR® necessary to freeze cells at 2×10^6 cells/vial:

$$\text{volume of mFreSR (ml)} = \frac{\text{Total cells counted}}{2 \times 10^6}$$

o. Aspirate supernatant and resuspend the pellet in the amount of mFreSR® calculated in 2.6.3n.

p. Dispense 1 ml mFreSR®/cell mixture into pre-labeled ampule for freezing and place on ice.

q. Freeze iPSC sample with the Control Rate Freezer (program number LCL1).

r. Transfer frozen ampules to liquid N_2 tank for long term storage.

·　·　·　·

C H A P T E R 3

General Protocols

3.1   EQUIPMENT LIST

1. Waterbath
   a) 37°C
   b) 56°C
2. Centrifuge
   a) Table-top—suitable for 15 and 50 ml tubes
   b) Microfuge
3. 37°C/5% CO_2 at 85% Relative humidity incubators
   a) 18% O_2
   b) 4% O_2
4. Level II biosafety cabinet
5. Argos Fireboy safety burner
6. Dissecting scope
7. Inverted phase contrast microscope
8. Inverted fluorescent microscope

3.2   MEDIA LIST
3.2.1 MEF Medium (500 ml)

   a) 440 ml high glucose DMEM (Life Technologies, catalog number 11995065)
   b) 50 ml Fetal Bovine Serum(Life Technologies, catalog number 10566016)
   c) 5 ml 100× L-glutamine (Life Technologies, catalog number 25030081)
   d) 5 ml non-essential amino acids (Life Technologies, catalog number 11140050)
   e) 0.9 ml 55 mM 2-Mercaptoethanol (Life Technologies, catalog number 21985023)

3.2.2 MEF Freezing Medium (100 ml)

   a. 10 ml MEF Medium
   b. 80 ml Fetal Bovine Serum (Life Technologies, catalog number 10439016)

    c.   10 ml Dimethylsulfoxide (DMSO) (Sigma-Aldrich, catalog number D2650-5X10ML)

3.2.3   Human Fibroblast Medium (500 ml)

    a.   435 ml High glucose DMEM (Life Technologies, catalog number 11995065)
    b.   50 ml FBS (Life Technologies, catalog number 10566016)
    c.   5 ml L-glutamine (Life Technologies, catalog number 25030081)
    d.   5 ml 100× Non-Essential Amino Acids (Life Technologies, catalog number 11140050)
    e.   5 ml 100× Pen/Strep (Life Technologies, catalog number 15070-063)

3.2.4   Human Fibroblast Enzyme Digestion Medium (100 ml)

    a.   Human fibroblast medium
    b.   1 g collagenase (Life Technologies, 17100-017)
    c.   1 unit/ml dispase (Life Technologies, 17105-041)

3.2.5   Human Fibroblast Freezing Medium (500 ml)

    a.   145 ml High glucose DMEM (Life Technologies, catalog number 11995065)
    b.   250 ml FBS (Life Technologies, catalog number 10566016)
    c.   100 ml DMSO (Sigma-Aldrich, catalog number D2650-5X10ML)
    d.   5 ml 100× Pen/Strep (Life Technologies, catalog number 15070-063)
    e.   5 ml 100× non-essential amino acids (Life Technologies, catalog number 11140050)

3.2.6   LB Broth (1000 ml)

    a.   10 g Bacto-tryptone
    b.   5 g yeast extract
    c.   10 g NaCl
    d.   Adjust pH to 7.5 with NaOH.
    e.   Adjust volume to 1 l with dH_2O.
    f.   Sterilize by autoclaving.

3.2.7   Plat A Cell Medium (250 ml)

    a.   435 ml high glucose DMEM (Life Technologies, catalog number 11995065)
    b.   50 ml Fetal Bovine Serum (Life Technologies, catalog number 10566016)

    c.   5 ml 100× Pen/Strep (Life Technologies, catalog number 15070-063)

    d.   5 ml 100× L-glutamine (Life Technologies, catalog number 25030081)

    e.   5 ml non-essential amino acids (Life Technologies, catalog number 11140050)

    f.   1 µg/ml puromycin (Life Technologies, catalog number A1113802)

    g.   10 µg/ml blasticidin (Life Technologies, catalog number A1113902)

3.2.8   Plat A Freezing Medium (50 ml)

    a.   10 ml <u>Plat A Cell Medium</u>

    b.   80 ml Fetal Bovine Serum(Life Technologies, catalog number 10566016)

    c.   10 ml DMSO (Sigma-Aldrich, catalog number D2650-5X10ML)

3.2.9   KOSR (250 ml)

    a.   195 ml DMEM/F12 (Life Technologies, catalog number 1210565018)

    b.   50 ml KnockOut™ Serum Replacement (Life Technologies, catalog number 10828028)

    c.   2.5 ml non-essential amino acids (Life Technologies, catalog number 11140050)

    d.   2.5 ml 100× L-glutamine (Life Technologies, catalog number 25030081)

    e.   50 µl 100 µg/ml Fibroblast Growth Factor (Peprotech, catalog number AF-100-18B)

3.2.10 mTeSR®1 (500 ml)

    a.   mTeSR®1 Supplement (Stem Cell Technologies, catalog number 05850)

    b.   mTeSR®1Basal Medium (Stem Cell Technologies, catalog number 05850)

    c.   Thaw and/or warm supplement and basal medium in 37°C water bath.

    d.   Add supplement to basal medium using 50 ml pipette (Ensure that the lot number on both the supplement and the basal medium are the same).

3.2.11 CPL RPMI Medium (250 ml)

    a.   207.5 ml Advanced RPMI (Life Technologies, catalog number 12633012)

    b.   37.5 ml Fetal Bovine Serum (Life Technologies, catalog number 10566016)

    c.   2.5 ml 100× L-Glutamine (Life Technologies, catalog number 25030081)

    d.   2.5 ml 100× Pen/Strep (Life Technologies, catalog number 15070-063)

3.2.12 CPL Freeze Medium (100 ml)

a. 40 ml CPL RPMI

b. 10 ml Dimethylsulfoxide (DMSO) (Sigma-Aldrich, catalog number D2650-5X10ML)

c. 50 ml Fetal Bovine Serum (Life Technologies, catalog number 10566016)

3.2.13 CPL Isolation/Activation Buffer

a. PBS

b. 0.1% Bovine Serum Albumin (Sigma-Aldrich, catalog number A2153)

c. 2 mM EDTA

3.3 PREPARATION OF GELATIN-COATED PLATES

Purpose: To coat culture dishes with gelatin for the growth of mEFs.

Materials:

1. 0.1% Gelatin

Procedure:

1. Add gelatin solution to plates according to chart below:

PLATE FORMAT	AMOUNT OF MG (mL)
96 well	0.1
48 well	0.25
24 well	0.5
12 well	0.75
6 well	1.0
Chamber slide	0.25
T25	3
T75	9
T175	15
10 cm	8

3.4   PREPARATION OF BD MATRIGEL™-COATED PLATES

Purpose: To coat culture dishes with Matrigel for the growth of pluripotent stem cells.

Materials:
1.   BD Matrigel™ (BD Biosciences, catalog number 354277)
2.   Ice cold DMEM/F12 (Life Technologies, catalog number 1210565018)

3.4.1   Thawing BD Matrigel™

Note: BD Matrigel™ is a solid at room temperature and should be thawed at 4°C or on ice (note— thawing may take up to 6 hours).

3.4.2   Aliquoting BD Matrigel™

Aliquot BD Matrigel™ in microfuge tubes at the concentration specified on the product specification sheet.

3.4.3   BD Matrigel™ Coating

    a.   Thaw aliquot of BD Matrigel™ on ice.

    b.   Resuspend BD Matrigel™ aliquot in 25 ml ice cold DMEM/F12. Keep diluted BD Matrigel™ tube on ice until use. Unused diluted BD Matrigel™ can be kept at 4°C for up to two weeks.

PLATE FORMAT	AMOUNT OF BD MATRIGEL™ (ML)
96 well	0.1
48 well	0.25
24 well	0.5
12 well	0.75
6 well	1.0
Chamber slide	0.25
T25	3
T75	9
T175	15
10 cm	8

c.   Add BD Matrigel™ solution to plates according to chart below.

d.   Incubate the BD Matrigel™-coated plate for at 37°C for at least 1 hour before use.

e.   Remove BD Matrigel™ and replace with mTeSR®1 or KOSR before plating iPSC.

3.5   HEMOCYTOMETER COUNTING

Purpose: Determine the density of a single cell suspension.

Materials:

1.   Trypan blue
2.   Hemocytometer

Procedure:

1.   Obtain a single cell suspension of the cells to be counted.
2.   Mix 90 µl of cell suspension and 10 µl trypan blue.
3.   Load hemocytometer with 10 µl of cell suspension.
4.   Count 4 corner squares (see Figure 11), but exclude any dead cells (blue from trypan blue uptake).

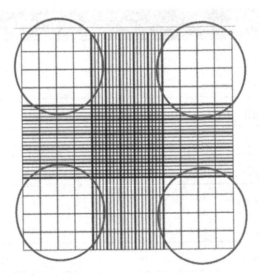

FIGURE 11: Hemocytometer. The squares used to obtain a cell count on a hemocytometer are circled in red.

5. $\dfrac{\text{total \# counted}}{4} * 10,000 = \dfrac{\text{cells}}{\text{ml}}$

6. $\dfrac{\text{Cells}}{\text{ml}} *$ total volume (ml) = total number of cells

3.6   MANUAL SCRAPING TOOL

Purpose: Generate a curved tool that can be used to remove areas of differentiation from an iPSC culture.

Materials:

1.   9½ in. sterile Pasteur pipettes

Procedure:

1.   Flame pasture pipette near where the barrel starts to narrow (Figure 12).

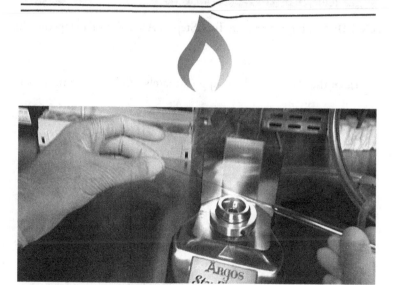

FIGURE 12 A AND B: Making a Scraping Tool Step 1. Flame the pipet where the barrel narrows to soften the pipet and bend into a 90°C angle.

2. As soon as the glass pipette start to melt and you sense the softness; quickly bend at a 90° angle. The result should be an "L" shape (Figure 13).

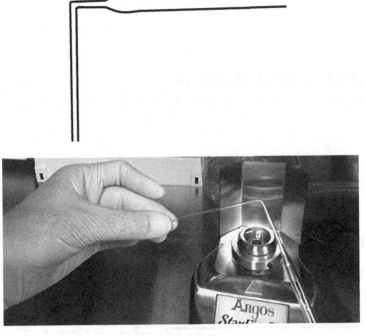

FIGURE 13 A AND B: Making a Scraping Tool Step 2. After the first step you should have an "L" shaped pipette.

3. Place the base of the "L" in the flame approximately 1/3 length from the stem (Figure 14).

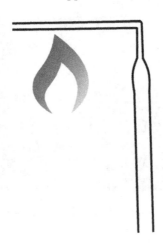

FIGURE 14 A AND B: Making a Scraping Tool Step 3. Place the pipet in the flame 1/3 of the way from the end of the "L" and apply constant pressure. As soon as the glass begins to melt and the pipet lengthens from the pressure, withdraw the pipet from the flame.

4. As soon as the glass starts to melt, pull the pipet in opposite directions to separate the pipet into 2 pieces. Note—the easiest way to achieve a scraping tool that has a nice end is to remove the pipet out of the flame as soon as it starts to pull apart. Once cooled, break at the thinnest part so that the pipette now has a very narrow end (Figure 15).

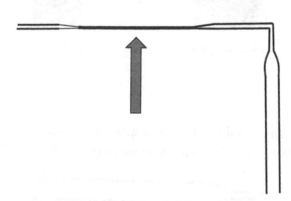

FIGURE 15: Making a Scraping Tool Step 4. Break pipet at the narrowest part of the drawn out region.

5. Flame the end of the tool to make a round ball which will be used for scraping differentiated cells (Figure 16).

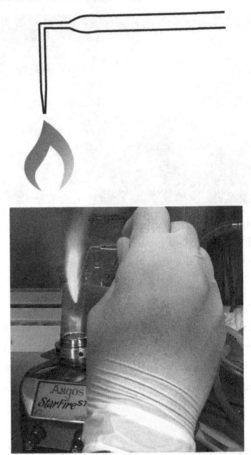

FIGURE 16 A AND B: Making a Scraping Tool Step 5. Invert the tip of the pipet in the flame to make a sealed rounded end.

6.   Resulting tool can be used to scrape away spontaneous differentiation in iPSC cultures or to manually select iPSC colonies for continued passage (Figure 17A and B).

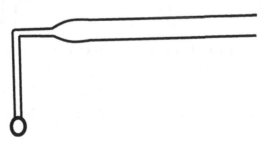

FIGURE 17A AND B: Complete Scraping Tool. The final scraping tool.

3.7 IDENTIFICATION AND MANUAL REMOVAL OF DIFFERENTIATION

Purpose: Remove spontaneous differentiation from an iPSC culture before passage.

Materials:

1. Scraping tool (made in "Manual Scraping Tool" protocol)
2. mTeSR®1

Procedure:

1. If there is a significant amount of cell debris change mTeSR®1 or KOSR so that colonies are more visible.

2. Inspect cultures to identify spontaneously differentiated cells.

   a. Pluripotent cells (shown in Figure 18):

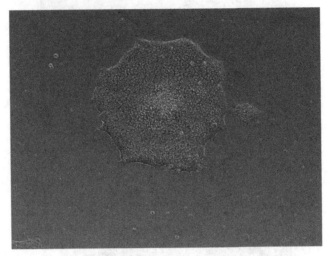

FIGURE 18: Morphology of a Pluripotent Colony. An example of the typical morphology of a pluripotent colony.

   i.   Compact, uniform colonies
   ii. Large nuclear to cytoplasm ratio
   iii. Slightly raised appearance

   b. Differentiated cells (shown in Figure 19):

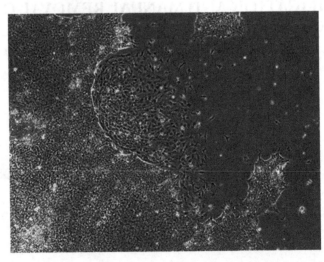

FIGURE 19: Morphology of Differentiation. An example of the Pluripotent Colony that has undergone spontaneous differentiation.

      i.   Flat, large cells

     ii.   Clear difference between nucleus and cytoplasm

   iii.   Irregular shaped non-compact cells

3. Holding the barrel end of the scraping tool, gently scrape off the differentiated cells.

Note: Plate should be kept at room temperature for no longer than 15 minutes for scraping. If more scraping time is necessary, place the plate back in incubator for 5 minutes to warm up, and then continue scraping to remove differentiated cells.

4. Swirl plate and aspirate media from the wells
5. Replace mTeSR®1 or other solution necessary for downstream application.

3.8   LIVE ALKALINE PHOSPHATASE STAINING OF IPSC

Purpose: Use live alkaline/phosphatase staining to identify pluripotent colonies.

Materials:

1) DMEM/F12 (Life Technologies, catalog number 1210565018)
2) Alkaline Phosphatase Live Stain stock solution (Life Technologies, Catalog number A14353)
   a. Stock solution should not be subjected to freeze/thaw cycles, so make into 5 µl aliquots in microfuge tubes.
   b. Prepare master mixes in DMEM/F12 as shown in the table below.

Procedure:

1. Wash cells 3 × 3 minutes in warm DMEM/F12.
2. Prepare Alkaline Phosphatase Live Stain during this time (use within 15 minutes of preparation).
3. Add Alkaline Phosphatase Live Stain to the wells that were washed with DMEM/F12 according to the table below and incubate for 20–30 minutes.
4. Remove Alkaline Phosphatase Live Stain.
5. Wash well 2 × 5 minutes with DMEM/F12.
6. Add fresh DMEM/F12.
7. Visualize under fluorescent microscope.
8. After visualization replace DMEM/F12 with mTeSR®1 or KOSR.

CULTURE FORMAT	AMOUNT OF 500× ALKALINE PHOSPHATASE LIVE STAIN (μL)	AMOUNT OF DMEM/F12 (mL)
6 well plate	3	1.5
12 well plate	1	0.5
10 cm dish	12	6.0

3.9   DISPASE DISSOCIATION OF iPSC

Purpose: Dissociate iPSC into colony pieces by dispase.

Materials:
1. BD™ Dispase (BD Biosciences, catalog number 354235)
2. Dispase Working solution
    a. Thaw 50 unit/ml stock BD™ dispase at room temperature
    b. Dilute 1:50 in DMEM/F12. This working solution is used for one day and works best if kept warm.
    c. The 50 unit/ml stock dispase vial should be frozen and thawed 5 times and a check mark should be made on the cap each time it is thawed.
3. DMEM/F12 (Life Technologies, catalog number 1210565018)
4. mTeSR®1 or KOSR

Procedure:
1. Remove spent medium from iPSC well.
2. Add 1 ml of Dispase working solution.
3. Incubate at 37°C for 5 minutes.
4. Remove dispase working solution.
5. Rinse 2 × 2 ml DMEM/F12.
6. Add 2 ml/well of mTeSR®1 or KOSR.
7. Scrape cells with cell scraper.
8. Transfer detached cells to 15 ml tube.
9. Use 1,000 μl micropipette to gently break the colonies into small pieces.

3.10 ACCUTASE DISSOCIATION OF IPSC

Purpose: Dissociate iPSC into a single cell suspension.

Materials:
1. PBS
2. mTeSR®1 or KOSR
3. Accutase™ (Stem Cell Technologies, Catalog number 07920)

Procedure:
1. Remove spent medium from iPSC well.
2. Wash well 2 × 2 ml of PBS.
3. Add 1 ml of Accutase™ to each well.
4. Incubate at 37°C for 5 minutes.
5. Use 1,000 µl micropipette to gently break the colonies into a single cell suspension and transfer to 15 ml tube.
6. Add 9 ml of PBS and mix gently.
7. Remove 10 µl of cell suspension and count following the "Hemocytometer Cell Counting Protocol."
8. Spin the 15 ml tube at 1,000 rpm for 5 minutes.
9. Aspirate supernatant and resuspend the cell pellet in mTeSR®1 or KOSR.

3.11 HEAT INACTIVATION OF FETAL BOVINE SERUM

Purpose: Prepare heat inactivated fetal bovine serum for use in CPL isolation protocol.

Materials:
1. Fetal Bovine Serum (FBS) (Life Technologies, catalog number 10566016)
2. 37°C water bath
3. 56°C water bath

Procedure:
1. Thaw FBS in 37°C water bath.
2. Transfer FBS to 56°C water bath.
3. Incubate for 30 minutes with swirling every 10 minutes.
4. Aliquot 1 ml of warm FBS into 1 ml vial and store at −20°C.
5. Thaw 1 ml vial on ice when needed.

3.12   THAWING AND ALIQUOTING SENDAI VIRUS

Purpose: Prepare single use aliquots of the hOct3/4, hSox2, hKlf4 and hc-Myc Sendai viruses used to generate iPSC.

Materials:

1. Cytotune™—iPS Sendai Reprogramming Kit (Life Technologies Catalog number, A1378001)

Procedure:

1. Thaw virus aliquots on ice (Note: each aliquot contains 100 μl of virus).
2. Aliquot 5 μl into microfuge tubes keeping everything on ice as much as possible.
3. Freeze and store at −80°C.

.   .   .   .

References

Ananiev, G., Williams, E.C., Li, H., and Chang, Q. (2011). Isogenic pairs of wild type and mutant induced pluripotent stem cell (iPSC) lines from Rett syndrome patients as in vitro disease model. *PLoS One* 6:e25255.

Cheung, A.Y., Horvath, L.M., Carrel, L., and Ellis, J. (2012). X-chromosome inactivation in rett syndrome human induced pluripotent stem cells. *Frontiers in Psychiatry / Frontiers Research Foundation* 3:24.

Cheung, A.Y., Horvath, L.M., Grafodatskaya, D., Pasceri, P., Weksberg, R., Hotta, A., Carrel, L., and Ellis, J. (2011). Isolation of MECP2-null Rett Syndrome patient hiPS cells and isogenic controls through X-chromosome inactivation. *Hum Mol Genet* 20: pp. 2103–15.

Chiu, A.Y., and Rao, M.S. (2011). Cell-based therapy for neural disorders—anticipating challenges. *Neurotherapeutics: The Journal of the American Society for Experimental NeuroTherapeutics* 8: pp. 744–52.

Chou, B.K., Mali, P., Huang, X., Ye, Z., Dowey, S.N., Resar, L.M., Zou, C., Zhang, Y.A., Tong, J., and Cheng, L. (2011). Efficient human iPS cell derivation by a non-integrating plasmid from blood cells with unique epigenetic and gene expression signatures. *Cell Res* 3: pp. 518–29.

Derosa, B.A., Van Baaren, J.M., Dubey, G.K., Vance, J.M., Pericak-Vance, M.A., and Dykxhoorn, D.M. (2012). Derivation of autism spectrum disorder-specific induced pluripotent stem cells from peripheral blood mononuclear cells. *Neurosci Lett* 1: pp. 9–14.

Dolmetsch, R., and Geschwind, D.H. (2011). The human brain in a dish: the promise of iPSC-derived neurons. *Cell* 145: pp. 831–4.

Ebert, A.D., Yu, J., Rose, F.F., Jr., Mattis, V.B., Lorson, C.L., Thomson, J.A., and Svendsen, C.N. (2009). Induced pluripotent stem cells from a spinal muscular atrophy patient. *Nature* 457: pp. 277–80.

Farra, N., Zhang, W.B., Pasceri, P., Eubanks, J.H., Salter, M.W., and Ellis, J. (2012). Rett syndrome induced pluripotent stem cell-derived neurons reveal novel neurophysiological alterations. *Mol Psychiatry*. DOI:10.1038/mp.2011.180

Fernandez, T.V., Sanders, S.J., Yurkiewicz, I.R., Ercan-Sencicek, A.G., Kim, Y.S., Fishman, D.O., Raubeson, M.J., Song, Y., Yasuno, K., Ho, W.S., et al. (2012). Rare copy number variants in tourette syndrome disrupt genes in histaminergic pathways and overlap with autism. *Biol Psychiatry* 71: pp. 392–402.

Gore, A., Li, Z., Fung, H.L., Young, J.E., Agarwal, S., Antosiewicz-Bourget, J., Canto, I., Giorgetti, A., Israel, M.A., Kiskinis, E., et al. (2011). Somatic coding mutations in human induced pluripotent stem cells. *Nature* 471: pp. 63–7.

Hargus, G., Cooper, O., Deleidi, M., Levy, A., Lee, K., Marlow, E., Yow, A., Soldner, F., Hockemeyer, D., Hallett, P.J., et al. (2010). Differentiated Parkinson patient-derived induced pluripotent stem cells grow in the adult rodent brain and reduce motor asymmetry in Parkinsonian rats. *Proc Natl Acad Sci U S A* 107: pp. 15921–6.

Holmans, P., Weissman, M.M., Zubenko, G.S., Scheftner, W.A., Crowe, R.R., Depaulo, J.R., Jr., Knowles, J.A., Zubenko, W.N., Murphy-Eberenz, K., Marta, D.H., et al. (2007). Genetics of recurrent early-onset major depression (GenRED): final genome scan report. *Am J Psychiatry* 164: pp. 248–58.

Hussein, S.M., Batada, N.N., Vuoristo, S., Ching, R.W., Autio, R., Narva, E., Ng, S., Sourour, M., Hamalainen, R., Olsson, C., et al. (2011). Copy number variation and selection during reprogramming to pluripotency. *Nature* 471: pp. 58–62.

Inoue, H., and Yamanaka, S. (2011). The use of induced pluripotent stem cells in drug development. *Clinical Pharmacology and Therapeutics* 89: pp. 655–61.

Kim, H.J., and Jin, C.Y. (2012). Stem cells in drug screening for neurodegenerative disease. *The Korean Journal of Physiology & Pharmacology: Official Journal of the Korean Physiological Society and the Korean Society of Pharmacology* 16: pp. 1–9.

Kim, K., Doi, A., Wen, B., Ng, K., Zhao, R., Cahan, P., Kim, J., Aryee, M.J., Ji, H., Ehrlich, L.I., et al. (2010). Epigenetic memory in induced pluripotent stem cells. *Nature* 467: pp. 285–90.

Larkin, J.E., Frank, B.C., Gavras, H., Sultana, R., and Quackenbush, J. (2005). Independence and reproducibility across microarray platforms. *Nature Methods* 2: pp. 337–44.

Lee, G., Papapetrou, E.P., Kim, H., Chambers, S.M., Tomishima, M.J., Fasano, C.A., Ganat, Y.M., Menon, J., Shimizu, F., Viale, A., et al. (2009). Modelling pathogenesis and treatment of familial dysautonomia using patient-specific iPSCs. *Nature* 461: pp. 402–6.

Liu, X., Li, F., Stubblefield, E.A., Blanchard, B., Richards, T.L., Larson, G.A., He, Y., Huang, Q., Tan, A.C., Zhang, D., et al. (2011). Direct reprogramming of human fibroblasts into dopaminergic neuron-like cells. *Cell Res* 22: pp. 321–332.

Loh, Y.H., Hartung, O., Li, H., Guo, C., Sahalie, J.M., Manos, P.D., Urbach, A., Heffner, G.C., Grskovic, M., Vigneault, F., et al. (2010). Reprogramming of T cells from human peripheral blood. *Cell Stem Cell* 7: pp. 15–9.

Marchetto, M.C., Carromeu, C., Acab, A., Yu, D., Yeo, G.W., Mu, Y., Chen, G., Gage, F.H., and Muotri, A.R. (2010). A model for neural development and treatment of Rett syndrome using human induced pluripotent stem cells. *Cell* 143: pp. 527–39.

McKay, R.D. (2004). Stem cell biology and neurodegenerative disease. *Philosophical Transactions of the Royal Society of London* 359: pp. 851–6.

Melas, P.A., Rogdaki, M., Osby, U., Schalling, M., Lavebratt, C., and Ekstrom, T.J. (2012). Epigenetic aberrations in leukocytes of patients with schizophrenia: association of global DNA methylation with antipsychotic drug treatment and disease onset. *FASEB J* 6: 2712–8.

Moore, J.C., Sadowy, S., Alikani, M., Toro-Ramos, A.J., Swerdel, M.R., Hart, R.P., and Cohen, R.I. (2010). A high-resolution molecular-based panel of assays for identification and characterization of human embryonic stem cell lines. *Stem Cell Res* 4: pp. 92–106.

Nazor, K.L., Altun, G., Lynch, C., Tran, H., Harness, J.V., Slavin, I., Garitaonandia, I., Muller, F.J., Wang, Y.C., Boscolo, F.S., et al. (2012). Recurrent variations in DNA methylation in human pluripotent stem cells and their differentiated derivatives. *Cell Stem Cell* 10; pp. 620–34.

Okubo, M., Tsurukubo, Y., Higaki, T., Kawabe, T., Goto, M., Murase, T., Ide, T., Furuichi, Y., and Sugimoto, M. (2001). Clonal chromosomal aberrations accompanied by strong telomerase activity in immortalization of human B-lymphoblastoid cell lines transformed by Epstein-Barr virus. *Cancer Genetics and Cytogenetics* 129: pp. 30–4.

Panchision, D.M. (2009). The role of oxygen in regulating neural stem cells in development and disease. *J Cell Physiol* 220: pp. 562–8.

Park, I.H., Zhao, R., West, J.A., Yabuuchi, A., Huo, H., Ince, T.A., Lerou, P.H., Lensch, M.W., and Daley, G.Q. (2008). Reprogramming of human somatic cells to pluripotency with defined factors. *Nature* 451: pp. 141–6.

Pasca, S.P., Portmann, T., Voineagu, I., Yazawa, M., Shcheglovitov, A., Pasca, A.M., Cord, B., Palmer, T.D., Chikahisa, S., Nishino, S., et al. (2011). Using iPSC-derived neurons to uncover cellular phenotypes associated with Timothy syndrome. *Nat Med* 17: pp. 1657–62.

Pedrosa, E., Sandler, V., Shah, A., Carroll, R., Chang, C., Rockowitz, S., Guo, X., Zheng, D., and Lachman, H.M. (2011). Development of patient-specific neurons in schizophrenia using induced pluripotent stem cells. *J Neurogenet* 25: pp. 88–103.

Rao, M.S., and Malik, N. (2012). Assessing iPSC reprogramming methods for their suitability in translational medicine. *J Cell Biochem.* DOI:10.1002/jcb.24183

Ripke, S., Sanders, A.R., Kendler, K.S., Levinson, D.F., Sklar, P., Holmans, P.A., Lin, D.Y., Duan, J., Ophoff, R.A., Andreassen, O.A., et al. (2011). Genome-wide association study identifies five new schizophrenia loci. *Nat Genet* 43: pp. 969–76.

Sanders, S.J., Ercan-Sencicek, A.G., Hus, V., Luo, R., Murtha, M.T., Moreno-De-Luca, D., Chu,

S.H., Moreau, M.P., Gupta, A.R., Thomson, S.A., et al. (2011). Multiple recurrent de novo CNVs, including duplications of the 7q11.23 Williams syndrome region, are strongly associated with autism. *Neuron* 70: pp. 863–85.

Seki, T., Yuasa, S., Oda, M., Egashira, T., Yae, K., Kusumoto, D., Nakata, H., Tohyama, S., Hashimoto, H., Kodaira, M., et al. (2010). Generation of induced pluripotent stem cells from human terminally differentiated circulating T cells. *Cell Stem Cell* 7: pp. 11–4.

Singh, U., Quintanilla, R.H., Grecian, S., Gee, K.R., Rao, M.S., and Lakshmipathy, U. (2012). Novel live alkaline phosphatase substrate for identification of pluripotent stem cells. *Stem Cell Reviews*. DOI:10.1007/s12015-012-9359-6

Sklar, P., Ripke, S., Scott, L.J., Andreassen, O.A., Cichon, S., Craddock, N., Edenberg, H.J., Nurnberger, J.I., Jr., Rietschel, M., Blackwood, D., et al. (2011). Large-scale genome-wide association analysis of bipolar disorder identifies a new susceptibility locus near ODZ4. *Nat Genet* 43: pp. 977–83.

Slaugenhaupt, S.A., Mull, J., Leyne, M., Cuajungco, M.P., Gill, S.P., Hims, M.M., Quintero, F., Axelrod, F.B., and Gusella, J.F. (2004). Rescue of a human mRNA splicing defect by the plant cytokinin kinetin. *Hum Mol Genet* 13: pp. 429–36.

Staerk, J., Dawlaty, M.M., Gao, Q., Maetzel, D., Hanna, J., Sommer, C.A., Mostoslavsky, G., and Jaenisch, R. (2010). Reprogramming of human peripheral blood cells to induced pluripotent stem cells. *Cell Stem Cell* 7: pp. 20–4.

Takahashi, K., Tanabe, K., Ohnuki, M., Narita, M., Ichisaka, T., Tomoda, K., and Yamanaka, S. (2007). Induction of pluripotent stem cells from adult human fibroblasts by defined factors. *Cell* 131: pp. 861–72.

Vaccarino, F.M., Stevens, H.E., Kocabas, A., Palejev, D., Szekely, A., Grigorenko, E.L., and Weissman, S. (2011). Induced pluripotent stem cells: a new tool to confront the challenge of neuropsychiatric disorders. *Neuropharmacology* 60: pp. 1355–63.

Wernig, M., Zhao, J.P., Pruszak, J., Hedlund, E., Fu, D., Soldner, F., Broccoli, V., Constantine-Paton, M., Isacson, O., and Jaenisch, R. (2008). Neurons derived from reprogrammed fibroblasts functionally integrate into the fetal brain and improve symptoms of rats with Parkinson's disease. *Proc Natl Acad Sci U S A* 105: pp. 5856–61.

Yu, J., Vodyanik, M.A., Smuga-Otto, K., Antosiewicz-Bourget, J., Frane, J.L., Tian, S., Nie, J., Jonsdottir, G.A., Ruotti, V., Stewart, R., et al. (2007). Induced pluripotent stem cell lines derived from human somatic cells. *Science* 318: pp. 1917–20.

Author Biographies

Dr. Jennifer Moore is the Associate Director of the National Institute of Mental Health's (NIMH) Stem Cell Resource at Rutgers, the State University of New Jersey. Dr. Moore received her Bachelor's of Science degree from the University of North Carolina at Charlotte in Chemistry and her Ph.D. from the University of North Carolina at Chapel Hill in Biochemistry and Biophysics. At the NIMH Stem Cell Resource, Dr. Moore oversees the generation of iPSC and their differentiation into neural derivatives for the study of psychiatric disorders.

Dr. Michael Sheldon is the Director of the National Institute of Mental Health (NIMH) Stem Cell Resource at the Rutgers University Cell and DNA Repository (RUCDR), the largest academically based cell and DNA repository in the United States. Dr. Sheldon received his Bachelor of Arts degree from Cornell University, and a Ph.D. in Genetics from SUNY at Stony Brook. Dr. Sheldon oversees all operational aspects of the NIMH Stem Cell Resource, including the processing of tissue biopsies for cultivation of primary source cells for iPSC reprogramming.

Dr. Ronald Hart is a professor in the department of Cell Biology & Neuroscience at Rutgers University and also a member of the W.M. Keck Center for Collaborative Neuroscience, the Rutgers Stem Cell Research Center, and the Human Genetics Institute of New Jersey. He obtained a B.S. from the University of Connecticut and a Ph.D. from the University of Michigan Medical School. His postdoctoral training was at Rockefeller University. His group studies molecular mechanisms of stem cell differentiation into neurons, focusing on the roles of non-coding RNAs, specific transcription factors, and epigenetic signaling.